MATHSWISE

Book One

Ray Allan
Head of Lower School,
Henry Compton School, Fulham.

Martin Williams
Head of Middle School,
Henry Compton School, Fulham.

Oxford University Press

...ord University Press, V...... Street, Oxford OX2 6DP

Oxford   New York   Toronto
Delhi   Bombay   Calcutta   Madras   Karachi
Petaling   Jaya   Singapore   Hong Kong   Tokyo
Nairobi   Dar es Salaam   Cape Town
Melbourne   Auckland

and associated companies in
Berlin   Ibadan

Oxford is a trademark of Oxford University Press

First Published 1985

First published 1985

Reprinted 1987, 1988

The authors and the publisher are grateful to Guinness
Superlatives Ltd for information obtained on various world
records which are inserted in this book. Readers are invited
to consult the current edition of *The Guinness Book of Records*
for the latest information on these and other records.

Cover illustration by Terry Pastor
Texts illustrations by Jon Riley

ISBN 0 19 834767 7

Artwork and Typesetting by BBG Ltd., Bristol.
Printed in Italy

Contents

Measurement

This work is about measuring and drawing accurately. It is important that you set out your work neatly. The drawing below shows you one way to set out your pages.

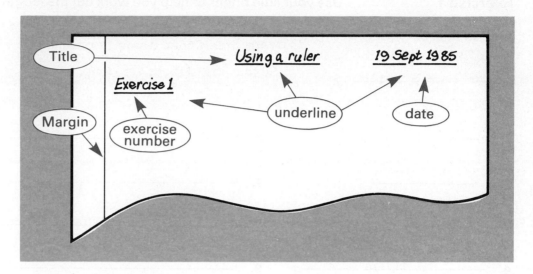

Things to remember

1. Use a pen for writing and a pencil for drawing.

2. If you make a mistake in pencil, rub it out.

3. If you make a mistake in ink, just put a neat line through the error and start again.

4. Read the questions carefully.

5. If you do not understand, always ask your teacher.

6. Use coloured pencils to brighten up your work.

7. Enjoy your maths work and try your best.

Measuring in centimetres

A centimetre is a unit of measurement.
A centimetre is about the width of your little finger.

Exercise 1

Use your little finger to help you work out the length of these lines.

Like this:

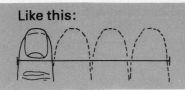

Your little finger would fit on this line about four times.

So this line is about 4 cm long.

1. ├──┤

2. ├────────┤

3. ├──────┤

4. ├──────────┤

5. ├────────┤

6. ├───────────────────────────┤

7. ├───────────────────┤

8. ├─────────────────────┤

9. ├──────────────────────────────┤

10. ├─────────────────────────┤

Exercise 2

These ten lines have been drawn again.

This time they have been marked in 1 cm sections.
Find the length of each line. See how close your guesses were.

Like this: 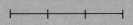 This line is 3 cm long.

1. ├──┤

2. ├──┼──┼──┤

3. ├──┼──┤

4. ├──┼──┼──┼──┼──┤

5. ├──┼──┼──┼──┤

6. ├──┼──┼──┼──┼──┼──┼──┼──┼──┤

7. ├──┼──┼──┼──┼──┼──┤

8. ├──┼──┼──┼──┼──┼──┼──┤

9. ├──┼──┼──┼──┼──┼──┼──┼──┼──┤

10. ├──┼──┼──┼──┼──┼──┤

Using a ruler

To measure a line you must start with the zero mark at the beginning of the line.

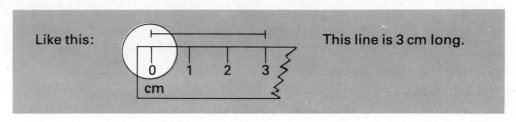

Like this: This line is 3 cm long.

Exercise 3   Write down the lengths of these lines in a sentence like the one above.

1.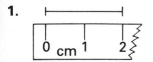

6.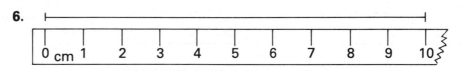

2.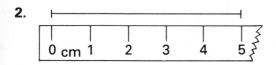

7.

3.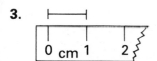

8.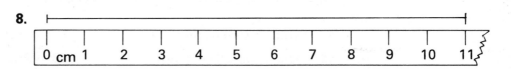

4.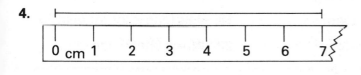

9.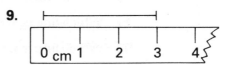

5.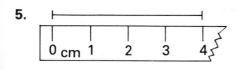

10.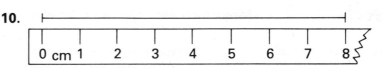

11.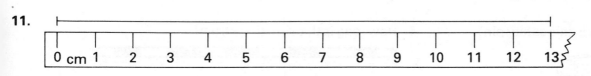

12.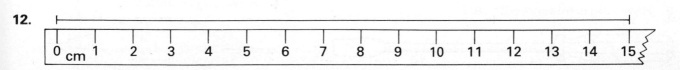

Exercise 4   Measure these lines to the nearest whole centimetre.

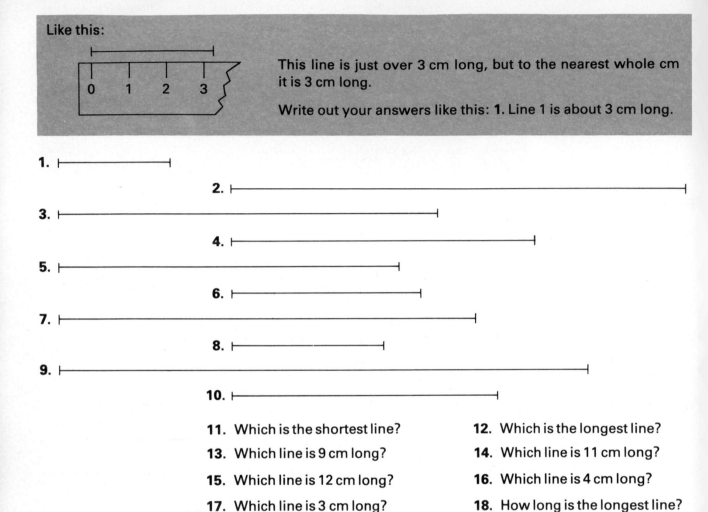
1. |———————|

2. |——————————————————————————————|

3. |————————————————|

4. |——————————|

5. |———————————————|

6. |——————————|

7. |————————————|

8. |—————————|

9. |———————————————————————————|

10. |————————————————|

11. Which is the shortest line?
12. Which is the longest line?
13. Which line is 9 cm long?
14. Which line is 11 cm long?
15. Which line is 12 cm long?
16. Which line is 4 cm long?
17. Which line is 3 cm long?
18. How long is the longest line?
19. Which line is 10 cm long?
20. Which line is 5 cm long?

Exercise 5   Measure these objects and say how long they are.

1. How long is this nail?

2. How long is this needle?

3. How long is this safety pin?

4. How long is this drinking straw?

5. How long is this screwdriver?

6. How long is this leaf?

Exercise 6   Measure these lines to the nearest half centimetre.

Like this

This ruler has half centimetre marks.
The line is about $3\frac{1}{2}$ cm long.

Write out your answers like this: **1.** Line 1 is $3\frac{1}{2}$ cm long.

1. ⊢————————⊣   **2.** ⊢——————————————⊣

3. ⊢———————————⊣   **4.** ⊢—————————————⊣

5. ⊢———⊣   **6.** ⊢———————————⊣

7. ⊢——————————⊣   **8.** ⊢—————————⊣

9. ⊢———————————⊣   **10.** ⊢——————⊣

11. ⊢———————————————⊣   **12.** ⊢—⊣

13. ⊢———————————————⊣

14. ⊢———————————————⊣

15. ⊢———————————————————⊣

Exercise 7   Copy out these sentences and fill in the blank spaces.

1. Line 'D' is ____ cm long.
2. Line 'H' is ____ cm long.
3. Line ____ is $5\frac{1}{2}$ cm long.
4. Line ____ and line ____ are both the same length.
5. Line 'C' is ____ cm long.
6. Line ____ is 5 cm long.
7. Line 'B' is ____ cm long.
8. Line ____ is the longest line.
9. Line 'G' is ____ cm long.
10. The shortest line is ____ cm long.

A   B   C   D   E   F   G   H

Exercise 8   Measure each object and complete each sentence.

1. This pencil is____cm long.

2. This key is____cm long.

3. This match is ____cm long.

4. This ink bottle is____cm wide and____cm high.

5. This bar of chocolate is____cm long and ____cm wide.

Choc Bar

6. This comb is ____cm long.

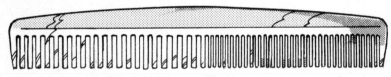

7. This doll is____cm high.

8. This pen knife is____cm long.

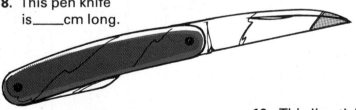

10. This lip-stick is ____cm long.

9. This screw is____cm long.

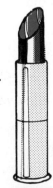

Here is a drawing of Robert Wadlow, the tallest person who ever lived.
He was 272 cm tall (8 feet 11 inches).

The shortest person was Pauline Musters.
She was only 59 cm tall (1 foot $11\frac{1}{2}$ inches).

If they walked into your classroom this is what they might look like.
If you wished to measure them, you would use a ruler or measuring tape.

Exercise 9

You will need a ruler and a metre of string.
Copy out the table below and complete it.

Myself.

1. I am ____cm tall.

2. The distance around my head is ____cm.

3. My arm is ____cm long.

4. The span of my hand is ____cm.

5. The distance around my wrist is ____cm.

6. The distance around my chest is ____cm.

7. The distance around my waist is ____cm.

8. The distance around my hips is____cm.

9. The distance between my knee and the floor is ____cm.

10. My shoe is ____cm long.

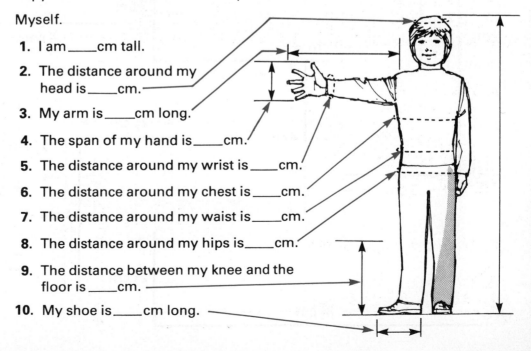

Drawing lines

In the next exercise you have to draw 20 lines. You should lay out your page like this:

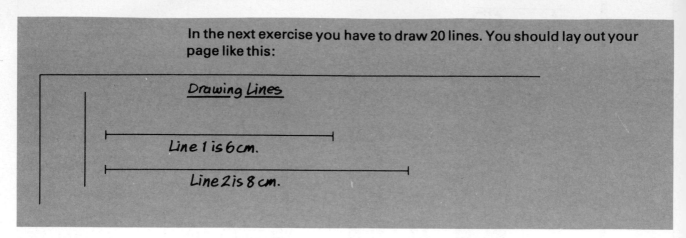

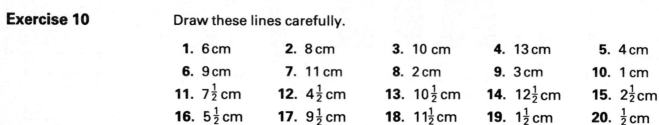

Exercise 10

Draw these lines carefully.

1. 6 cm	**2.** 8 cm	**3.** 10 cm	**4.** 13 cm	**5.** 4 cm
6. 9 cm	**7.** 11 cm	**8.** 2 cm	**9.** 3 cm	**10.** 1 cm
11. $7\frac{1}{2}$ cm	**12.** $4\frac{1}{2}$ cm	**13.** $10\frac{1}{2}$ cm	**14.** $12\frac{1}{2}$ cm	**15.** $2\frac{1}{2}$ cm
16. $5\frac{1}{2}$ cm	**17.** $9\frac{1}{2}$ cm	**18.** $11\frac{1}{2}$ cm	**19.** $1\frac{1}{2}$ cm	**20.** $\frac{1}{2}$ cm

Exercise 11

Use squared paper to draw these lines.

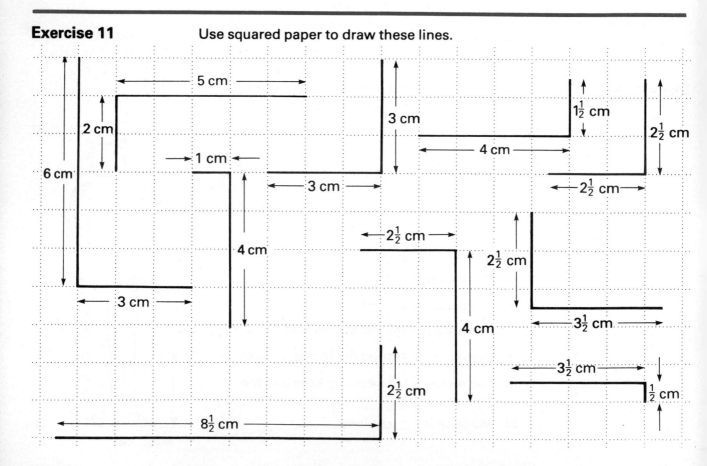

Exercise 12   Draw these shapes on squared paper.

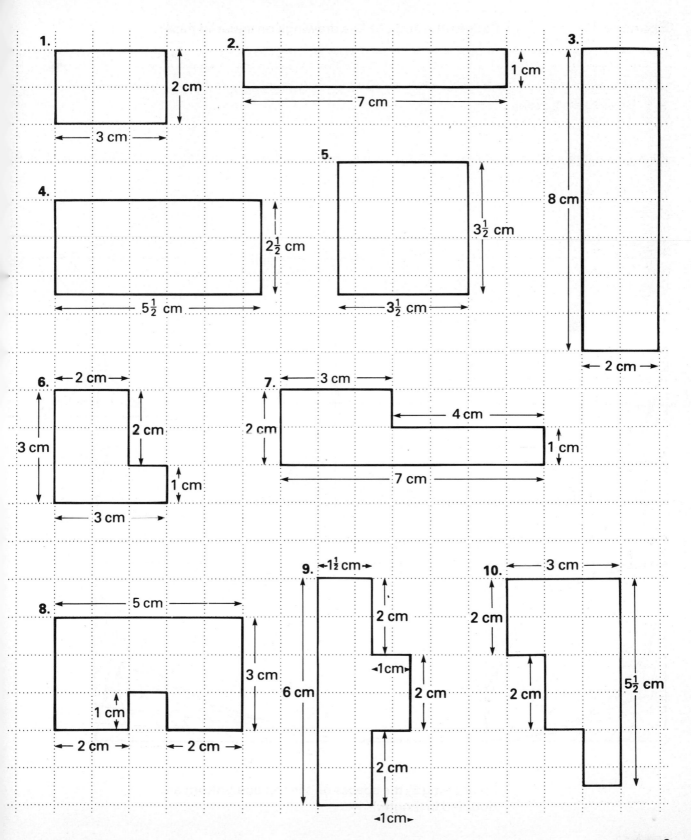

Making curves from straight lines

Exercise 13 Redraw the straight line drawings on squared paper.

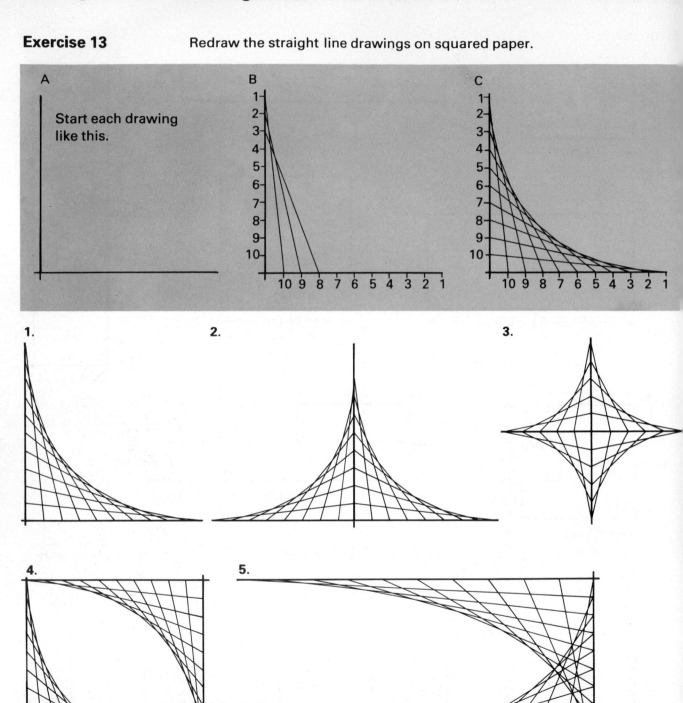

(In drawing 5, the spaces on the top and bottom are doubled in length.)

Sorting

Exercise 1   Use the head sorter to match these dummies' heads to their correct bodies.

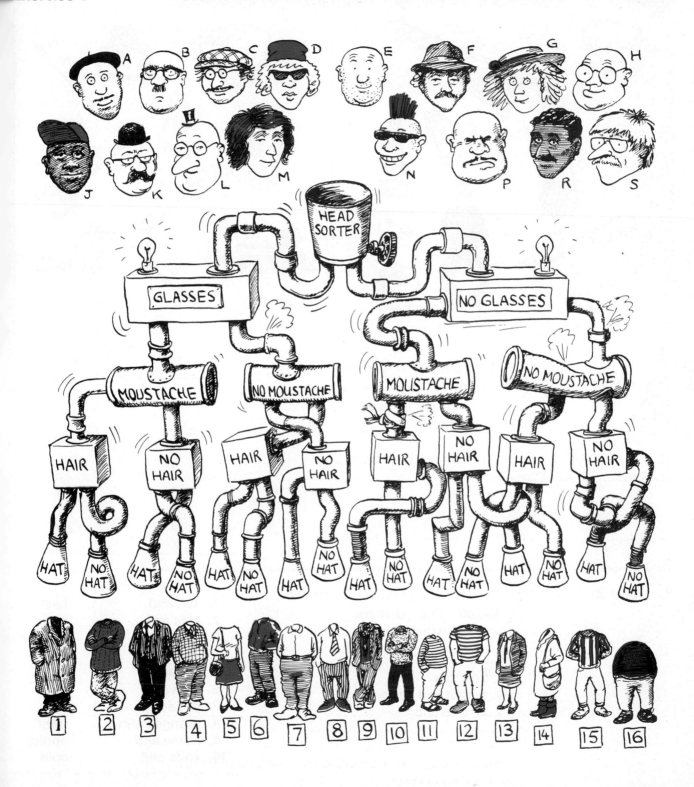

Exercise 2

Look at the objects in the box on the left.
Can you find their partners in the box on the right?

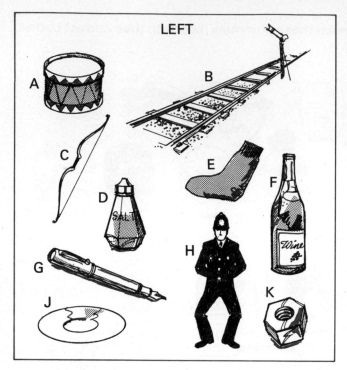

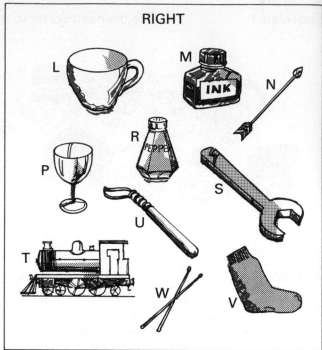

Copy out the table on the right.

Use arrows to join the letters of each
object on the left with its partner
on the right.

The first one is done for you.

LEFT	RIGHT
A	L
B	M
C	N
D	P
E	R
F	S
G	T
H	U
J	V
K	W

Exercise 3

Here is another table of partners.
Match up the partners
and write them out.

Like this:
1. salt and pepper
2. cats and

1.	salt and	bacon
2.	cats and	boys
3.	heads and	pepper
4.	egg and	cart
5.	left and	tails
6.	girls and	dogs
7.	hot and	fork
8.	oranges and	right
9.	horse and	apples
10.	knife and	cold

Sorting into sets

Peter has to tidy his bedroom. He has to put his things into drawers.

Exercise 4

Draw each of the drawers, like the ones below.
Put the letter of each object into the correct drawer.

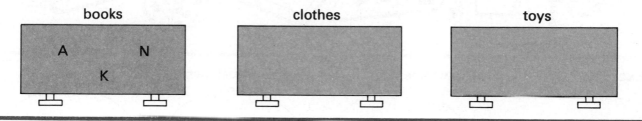

books   clothes   toys

A   N

K

Exercise 5

In the garage, Peter's father is sorting out his tools.
He puts them into three groups
1. Cutting tools.   **2.** Measuring tools.   **3.** Screwdrivers.

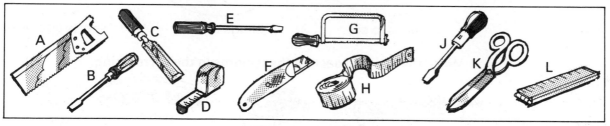

Sort the tools into three groups.
Draw three rings like the ones below.
Label them Cutting tools, Measuring tools and Screwdrivers.
Put the correct letter in each ring.

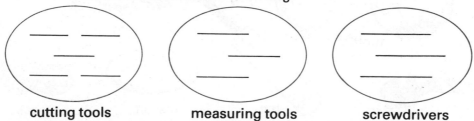

cutting tools   measuring tools   screwdrivers

Exercise 6   Draw three rings in your book and label them like this.

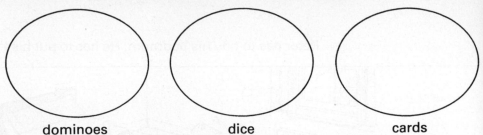

dominoes   dice   cards

Sort these items into three groups.   **1.** dominoes   **2.** dice   **3.** cards
Put the letter of each item into the correct ring.

Exercise 7   Draw three rings in your book and label them like this.

things to eat   things to wear   sports equipment

Write the name of each of these items in the correct ring.

sandwich   sweet   hat   sausage

football   egg   scarf   hockey stick

tie   bat   biscuit   sock

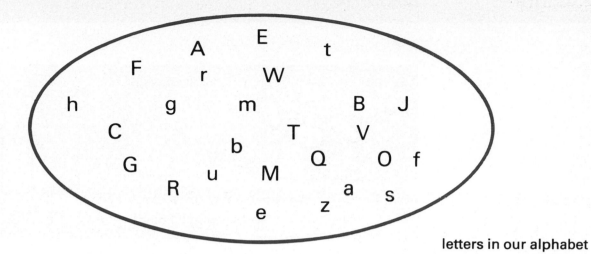

letters in our alphabet

Exercise 8

This is a group of letters from the alphabet. Some letters are small and some are capital letters.

1. Sort these letters into two groups:
   'small letters' and 'capital letters'.
   Put the letters into the correct ring.
   You can see how to do this on the right.

   capital letters   small letters

2. Now sort these letters into two different groups.
   Some letters are made from straight lines only,
   and some are made from curves.
   Draw two more rings and label them
   'straight line letters' and 'curved letters'
   Put all the letters into the correct ring.

Exercise 9

Draw five rings in your book and label them as shown below.

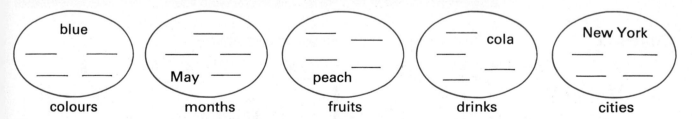

colours   months   fruits   drinks   cities

Look at the list below. Put each word in the correct ring.
Some have been done to show you how.

apple	December			green
cola	May	January	blue	
red	Paris	strawberry	Glasgow	grape
November	beer	banana	Rome	tea
London	peach	New York	brown	July
	yellow	water	coffee	

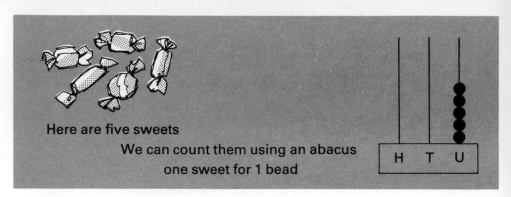

Here are five sweets

We can count them using an abacus
one sweet for 1 bead

Exercise 1

Draw five abacuses like the one above. Put the correct number of beads on the spike to count each group of objects.

1. 2. 3. 4. 5.

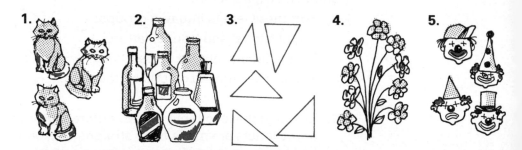

You can only put 9 beads on one spike.

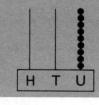

How many beads are shown here?
There are 9 beads on the unit spike.
We write the answer: 9 beads.

Exercise 2

Write down the number shown on each abacus.

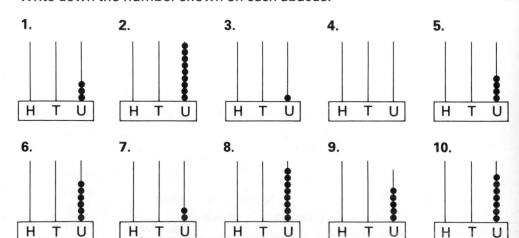

1. 2. 3. 4. 5.

6. 7. 8. 9. 10.

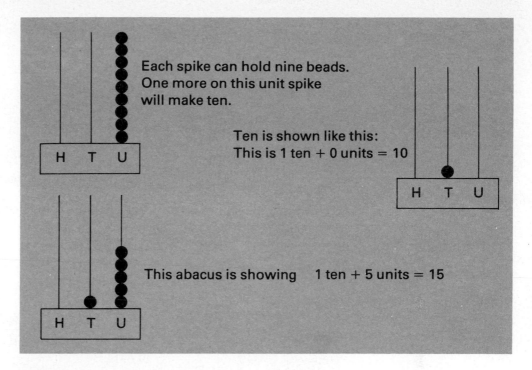

Each spike can hold nine beads.
One more on this unit spike
will make ten.

Ten is shown like this:
This is 1 ten + 0 units = 10

This abacus is showing   1 ten + 5 units = 15

Exercise 3

Write down the number shown on each abacus.

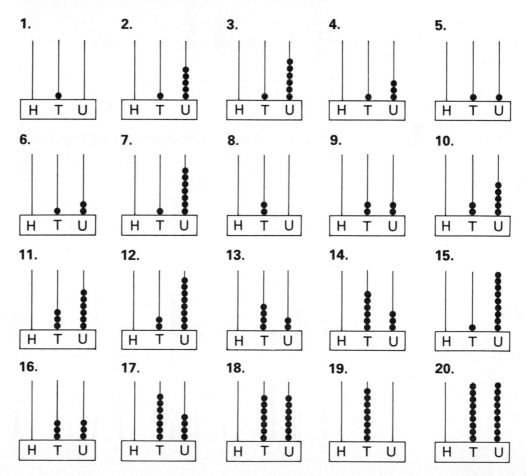

1.   2.   3.   4.   5.

6.   7.   8.   9.   10.

11.   12.   13.   14.   15.

16.   17.   18.   19.   20.

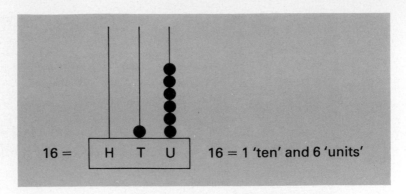

16 = | H | T | U | 16 = 1 'ten' and 6 'units'

Exercise 4

Show these numbers on abacuses like the one above.

1. 16	**2.** 19	**3.** 20	**4.** 24	**5.** 1?
6. 27	**7.** 29	**8.** 33	**9.** 61	**10.** 6?
11. 72	**12.** 53	**13.** 44	**14.** 91	**15.** 7?

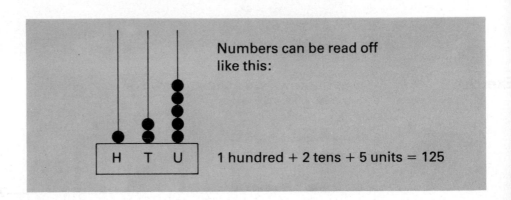

Numbers can be read off like this:

1 hundred + 2 tens + 5 units = 125

Exercise 5

Write down the number shown on each abacus.

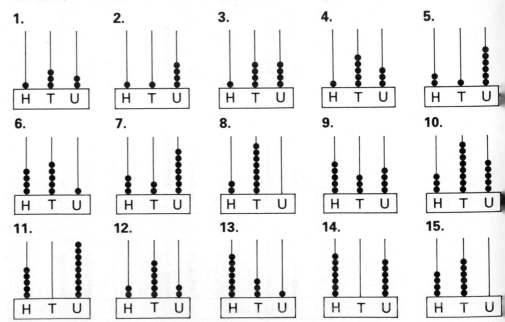

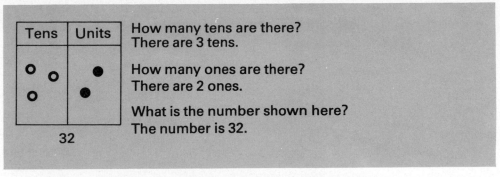

Tens	Units
o o o	● ●

32

How many tens are there?
There are 3 tens.

How many ones are there?
There are 2 ones.

What is the number shown here?
The number is 32.

Exercise 6

Answer these three questions for each of the boxes below.

How many tens are there?
How many ones are there?
What is the number shown in the box?

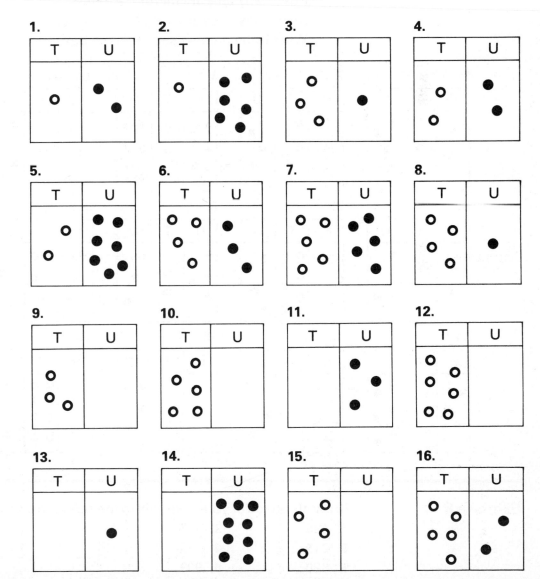

Hundreds	Tens	Units
▲ ▶	○ ○ ○ ○ ○	● ● ● ● ● ●
2	5	6

How many hundreds are there?
There are 2 hundreds.
How many tens are there?
There are 5 tens.
How many ones are there?
There are 6 ones.
What is the number shown here?
The number is 256.

Exercise 7

Answer these four questions for each of the boxes below.

    How many hundreds are there?
    How many tens are there?
    How many ones are there?
    What is the number shown in the box?

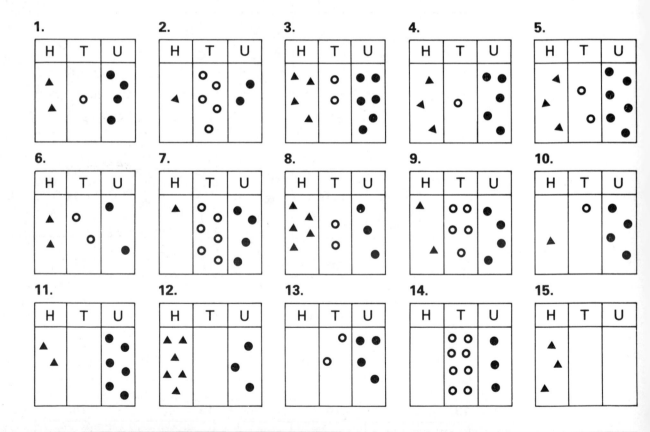

Exercise 8

Draw six boxes of your own. Fill in the boxes to show these numbers.

1. 425
2. 333
3. 214
4. 28
5. 320
6. 909

What happens for still bigger numbers?
We need another column.
This box shows the number 2000.
There are 2 thousands.

Th	H	T	U
△ △			
2	0	0	0

The position of the digits in a number
is very important.

Th	H	T	U
2	0	0	0

1562 The 2 stands for 2 units   1526 The 2 stands for 2 tens.
1256 The 2 stands for 2 hundreds   2156 The 2 stands for 2 thousands.
The digits further to the left stand for bigger and bigger numbers.

Exercise 9

Answer the questions about these numbers.

	Th	H	T	U	
1.		2	5	6	What does the 6 stand for?
2.			8	7	What does the 8 stand for?
3.	3	4	9	0	What does the 4 stand for?
4.	5	0	8	2	What does the 5 stand for?
5.	7	6	4	0	What does the 4 stand for?
6.	6	4	7	3	The 6 stands for 6 _____
7.		4	9	0	The 0 stands for 0 _____
8.		9	5	8	The 9 stands for 9 _____
9.	8	0	9	0	The 9 stands for 9 _____
10.		7	9	9	The 7 stands for 7 _____

Exercise 10

1. Which of these numbers has a 9 in the units place?
   a. 940   b. 490   c. 409

2. Which of these numbers has a 0 in the tens place?
   a. 509   b. 210   c. 2560

3. Which of these numbers has a 5 in the thousands place?
   a. 525   b. 5001   c. 355

4. Which of these numbers has a 7 in the hundreds place?
   a. 572   b. 117   c. 773

Carrying

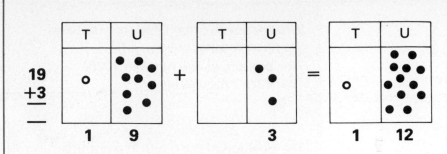

$$\begin{array}{r} 19 \\ +3 \\ \hline \end{array}$$

1   9   3   1   12

You can only have 9 units in the units column.
So 10 of the 12 units are changed to 1 ten.
Place the new ten in
the Tens column:

The number then
becomes:

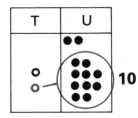

10

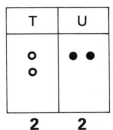

$$\begin{array}{r} 19 \\ +3 \\ \hline 22 \end{array}$$

2   2

Exercise 11

Draw out the boxes. Complete the boxes and answer the addition
questions.

1.

$$\begin{array}{r} 16 \\ +7 \\ \hline \end{array}$$

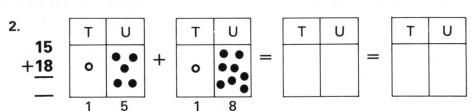

1   6   7   1   13

2.

$$\begin{array}{r} 15 \\ +18 \\ \hline \end{array}$$

1   5   1   8

3.

$$\begin{array}{r} 28 \\ +2 \\ \hline \end{array}$$

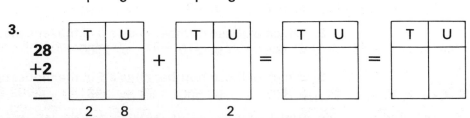

2   8   2

Draw out four boxes for each of these sums. Then copy and complete the
sum.

4. $12 + 9 =$ ___   **5.** $17 + 6 =$ ___   **6.** $14 + 6 =$ ___   **7.** $29 + 5 =$ ___

Exercise 1   Solve these problems. See how many you can do in 30 minutes.

1. There are _____ mushrooms here.

2. There are _____ apples here.

3. There are _____ sweets here.

4. There are _____ cubes here.

Copy out these sums. What number or sign is missing from each box?

5. 2
    +3
    ☐

6. 4
    +☐
    5

7. 6
    ☐3
    9

8. 4
    +0
    ☐

9. ☐
    +6
    8

10. 5
     ☐2
     7

11. 4
     ☐1
     5

12. 3
     +☐
     8

13. ☐
     ☐6
     9

14. 0
     ☐8
     8

15. 5
     +☐
     8

16. ☐
     +2
     3

17. 2
     +☐
     7

18. ☐
     +4
     4

19. 2 and 5 is _____   20. 5 and 4 is _____   21. 4 plus 3 is _____
22. 2 and 7 equals _____   23. 8 and 1 equals _____   24. 6 and 2 makes _____
25. 7 add 0 is _____   26. 4 plus 5 equals _____   27. 0 and 9 makes _____
28. 2 and 5 equals _____   29. 4 plus 4 equals _____   30. 3 and 6 makes _____

31. Which box has the most buttons?

A   B   C

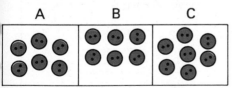

Box _____ has the most buttons.

32. Which box has the most ticks?

A   B   C

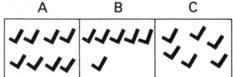

Box _____ has the most ticks.

33. Which box has the most crosses?

A   B   C

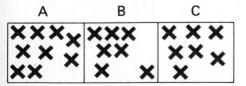

Box _____ has the most crosses.

34. Which box has the most dots?

A   B   C

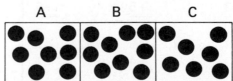

Box _____ has the most dots.

Exercise 2

Do these problems in your book.

1. 3 $+2$	**2.** 4 $+3$	**3.** 5 $+2$	**4.** 2 $+7$	**5.** 6 $+1$
6. 5 $+3$	**7.** 1 $+7$	**8.** 8 $+1$	**9.** 7 $+2$	**10.** 3 $+3$
11. 4 $+4$	**12.** 0 $+4$	**13.** 4 $+5$	**14.** 7 $+0$	**15.** 0 $+3$

Workcard 1

1. 8 $+1$	**2.** 4 $+3$	**3.** 5 $+2$
4. 6 $+0$	**5.** 3 $+5$	**6.** 4 $+4$
7. 6 $+2$	**8.** 1 $+6$	**9.** 0 $+7$
10. 2 $+3$	**11.** 5 $+1$	**12.** 4 $+5$

Workcard 2

1. 5 $+5$	**2.** 6 $+6$	**3.** 5 $+7$
4. 7 $+4$	**5.** 6 $+4$	**6.** 1 $+9$
7. 8 $+3$	**8.** 8 $+2$	**9.** 6 $+5$
10. 9 $+4$	**11.** 7 $+6$	**12.** 2 $+8$

Workcard 3

1. 6 $+3$	**2.** 8 $+1$	**3.** 7 $+0$
4. 10 $+2$	**5.** 10 $+3$	**6.** 10 $+7$
7. 11 $+3$	**8.** 11 $+5$	**9.** 12 $+5$
10. 13 $+3$	**11.** 15 $+4$	**12.** 17 $+1$

Workcard 4

1. 10 $+8$	**2.** 12 $+7$	**3.** 16 $+2$
4. 9 $+6$	**5.** 7 $+7$	**6.** 8 $+8$
7. 10 $+2$	**8.** 14 $+0$	**9.** 15 $+2$
10. 9 $+10$	**11.** 5 $+10$	**12.** 4 $+12$

Workcard 5

1. $\begin{array}{r} 4 \\ +9 \\ \hline \end{array}$	**2.** $\begin{array}{r} 6 \\ +0 \\ \hline \end{array}$	**3.** $\begin{array}{r} 5 \\ +5 \\ \hline \end{array}$
4. $\begin{array}{r} 9 \\ +1 \\ \hline \end{array}$	**5.** $\begin{array}{r} 6 \\ +7 \\ \hline \end{array}$	**6.** $\begin{array}{r} 8 \\ +3 \\ \hline \end{array}$
7. $\begin{array}{r} 14 \\ +4 \\ \hline \end{array}$	**8.** $\begin{array}{r} 6 \\ +13 \\ \hline \end{array}$	**9.** $\begin{array}{r} 11 \\ +6 \\ \hline \end{array}$
10. $\begin{array}{r} 2 \\ +17 \\ \hline \end{array}$	**11.** $\begin{array}{r} 0 \\ +18 \\ \hline \end{array}$	**12.** $\begin{array}{r} 6 \\ +10 \\ \hline \end{array}$

Workcard 6

1. $\begin{array}{r} 9 \\ +5 \\ \hline \end{array}$	**2.** $\begin{array}{r} 9 \\ +9 \\ \hline \end{array}$	**3.** $\begin{array}{r} 8 \\ +9 \\ \hline \end{array}$
4. $\begin{array}{r} 14 \\ +3 \\ \hline \end{array}$	**5.** $\begin{array}{r} 15 \\ +1 \\ \hline \end{array}$	**6.** $\begin{array}{r} 17 \\ +2 \\ \hline \end{array}$
7. $\begin{array}{r} 13 \\ +6 \\ \hline \end{array}$	**8.** $\begin{array}{r} 3 \\ +13 \\ \hline \end{array}$	**9.** $\begin{array}{r} 3 \\ +16 \\ \hline \end{array}$
10. $\begin{array}{r} 6 \\ +6 \\ \hline \end{array}$	**11.** $\begin{array}{r} 3 \\ +9 \\ \hline \end{array}$	**12.** $\begin{array}{r} 5 \\ +9 \\ \hline \end{array}$

Workcard 7

1. $\begin{array}{r} 10 \\ +11 \\ \hline \end{array}$	**2.** $\begin{array}{r} 14 \\ +10 \\ \hline \end{array}$	**3.** $\begin{array}{r} 13 \\ +15 \\ \hline \end{array}$
4. $\begin{array}{r} 12 \\ +15 \\ \hline \end{array}$	**5.** $\begin{array}{r} 16 \\ +13 \\ \hline \end{array}$	**6.** $\begin{array}{r} 13 \\ +13 \\ \hline \end{array}$
7. $\begin{array}{r} 18 \\ +11 \\ \hline \end{array}$	**8.** $\begin{array}{r} 15 \\ +11 \\ \hline \end{array}$	**9.** $\begin{array}{r} 14 \\ +14 \\ \hline \end{array}$
10. $\begin{array}{r} 21 \\ +11 \\ \hline \end{array}$	**11.** $\begin{array}{r} 22 \\ +14 \\ \hline \end{array}$	**12.** $\begin{array}{r} 16 \\ +23 \\ \hline \end{array}$

Workcard 8

1. $\begin{array}{r} 23 \\ +23 \\ \hline \end{array}$	**2.** $\begin{array}{r} 23 \\ +26 \\ \hline \end{array}$	**3.** $\begin{array}{r} 21 \\ +25 \\ \hline \end{array}$
4. $\begin{array}{r} 27 \\ +22 \\ \hline \end{array}$	**5.** $\begin{array}{r} 25 \\ +24 \\ \hline \end{array}$	**6.** $\begin{array}{r} 26 \\ +22 \\ \hline \end{array}$
7. $\begin{array}{r} 32 \\ +24 \\ \hline \end{array}$	**8.** $\begin{array}{r} 36 \\ +23 \\ \hline \end{array}$	**9.** $\begin{array}{r} 35 \\ +33 \\ \hline \end{array}$
10. $\begin{array}{r} 21 \\ +42 \\ \hline \end{array}$	**11.** $\begin{array}{r} 51 \\ +25 \\ \hline \end{array}$	**12.** $\begin{array}{r} 35 \\ +43 \\ \hline \end{array}$

Workcard 9

1. $\begin{array}{r} 7 \\ +13 \\ \hline \end{array}$	**2.** $\begin{array}{r} 6 \\ +14 \\ \hline \end{array}$	**3.** $\begin{array}{r} 5 \\ +15 \\ \hline \end{array}$
4. $\begin{array}{r} 16 \\ +5 \\ \hline \end{array}$	**5.** $\begin{array}{r} 13 \\ +8 \\ \hline \end{array}$	**6.** $\begin{array}{r} 15 \\ +8 \\ \hline \end{array}$
7. $\begin{array}{r} 18 \\ +4 \\ \hline \end{array}$	**8.** $\begin{array}{r} 14 \\ +7 \\ \hline \end{array}$	**9.** $\begin{array}{r} 12 \\ +9 \\ \hline \end{array}$
10. $\begin{array}{r} 6 \\ +15 \\ \hline \end{array}$	**11.** $\begin{array}{r} 4 \\ +19 \\ \hline \end{array}$	**12.** $\begin{array}{r} 7 \\ +16 \\ \hline \end{array}$

Workcard 10

1. $\begin{array}{r} 23 \\ +41 \\ \hline \end{array}$	**2.** $\begin{array}{r} 33 \\ +14 \\ \hline \end{array}$	**3.** $\begin{array}{r} 17 \\ +32 \\ \hline \end{array}$
4. $\begin{array}{r} 19 \\ +1 \\ \hline \end{array}$	**5.** $\begin{array}{r} 18 \\ +5 \\ \hline \end{array}$	**6.** $\begin{array}{r} 17 \\ +7 \\ \hline \end{array}$
7. $\begin{array}{r} 26 \\ +6 \\ \hline \end{array}$	**8.** $\begin{array}{r} 27 \\ +4 \\ \hline \end{array}$	**9.** $\begin{array}{r} 35 \\ +6 \\ \hline \end{array}$
10. $\begin{array}{r} 13 \\ +29 \\ \hline \end{array}$	**11.** $\begin{array}{r} 18 \\ +23 \\ \hline \end{array}$	**12.** $\begin{array}{r} 16 \\ +26 \\ \hline \end{array}$

Workcard 11

1. 34
  +8

2. 26
  +7

3. 35
  +16

4. 44
  +17

5. 36
  +16

6. 38
  +26

7. 35
  +17

8. 17
  +37

9. 24
  +18

10. 37
   +24

11. 35
   +36

12. 28
   +27

Workcard 12

1. 46
  +37

2. 29
  +39

3. 38
  +18

4. 44
  +37

5. 39
  +19

6. 28
  +15

7. 9
  +34

8. 7
  +27

9. 4
  +38

10. 58
   + 9

11. 46
   +17

12. 27
   +35

Exercise 3

Which two numbers from this circle add up to 12?
7 and 5 add up to 12.

1.

a. Which two numbers add up to 15?
b. Which two numbers add up to 10?
c. Which two numbers add up to 5?

2.

a. Which two numbers add up to 19?
b. Which two numbers add up to 14?
c. Which two numbers add up to 10?

3.
a. Which two numbers add up to 16?
b. Which two numbers add up to 12?
c. Which two numbers add up to 13?

4.
a. Which two numbers add up to 20?
b. Which two numbers add up to 16?
c. Which two numbers add up to 12?

5.
a. Which two numbers add up to 23?
b. Which two numbers add up to 32?
c. Which two numbers add up to 14?

6.

a. Which two numbers add up to 27?
b. Which two numbers add up to 18?
c. Which two numbers add up to 24?

This machine makes number sausages.
It puts numbers into sausages.
Each sausage is worth ⑥:

$$3 + 2 + 1 = 6$$
$$4 + 2 = 6$$
$$3 + 3 = 6$$
$$5 + 1 = 6$$

Exercise 4

Copy the numbers. Put a sausage around each group of numbers to make them add up to the number in red.

1. 2, 1, 3, 2, 4, 4, 5, 2, 1 . . . ⑧

2. 1, 1, 2, 1, 4, 1, 3, 2, 5, 0 . . . ⑤

3. 5, 5, 6, 4, 3, 2, 5 . . . ⑩

4. 1, 6, 5, 1, 1, 3, 2, 1, 1, . . . ⑦

5. 2, 2, 2, 1, 1, 4, 2, 1, 1, 2 . . . ⑥

6. 2, 3, 4, 1, 1, 1, 3, 2, 1, 2 . . . ⑤

7. 1, 2, 1, 2, 2, 1, 1, 1, 1, 3, 1 . . . ④

8. 2, 1, 1, 2, 1, 5, 3, 3, 2, 4 . . . ⑥

9. 7, 2, 3, 4, 2, 1, 6, 2, 1, 8 . . . ⑨

10. 7, 3, 4, 3, 3, 5, 1, 1, 1, 2, 2, 8 . . . ⑩

This machine has broken down.

The numbers come out in the wrong order.
They must be rearranged and then put into sausages.

2, 1, 1, 4, 3, 2, 2 . . . ⑤
The numbers should add up to five.

Exercise 5

Copy out the string of numbers.

Rearrange the numbers. Put sausages around the numbers to make them add up to the numbers in red.

1. 6, 2, 1, 1, 7, 5, 3, 2, 5 . . . ⑧

2. 1, 3, 5, 3, 4, 1, 4, 1, 5, 2, 1 . . . ⑥

3. 1, 1, 4, 3, 1, 4, 6, 5, 2, 1 . . . ⑦

4. 1, 1, 2, 3, 1, 1, 1, 3, 4, 3 . . . ⑤

5. 6, 1, 2, 4, 7, 4, 1, 1, 6, 2, 6 . . . ⑧

6. 4, 1, 1, 3, 7, 5, 6, 8, 8, 1, 1 . . . ⑨

7. 1, 1, 3, 2, 3, 1, 1, 1, 3, 1, 3 . . . ④

8. 3, 1, 3, 4, 2, 3, 1, 1, 4, 1, 2 . . . ⑤

Exercise 6

Match each sum with its equal partner in the ring: $2 + 3 = 1 + 4$

1. $2 + 3 =$

2. $1 + 6 =$

3. $2 + 2 =$

4. $5 + 3 =$

5. $4 + 5 =$

6. $1 + 2 =$

7. $5 + 5 =$

8. $6 + 7 =$

9. $7 + 5 =$

10. $8 + 6 =$

11. $7 + 8 =$

12. $6 + 5 =$

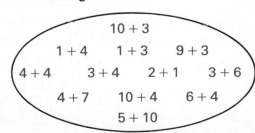

Exercise 7   Number squares

2	3	
5	1	
		☐

Add the numbers across.

2	3	5
5	1	6
		☐

Now add the numbers downward.

2	3	5
5	1	6
7	4	☐

Now add the answers across and then down.
The answer should be the same both ways.

2	3	5
5	1	6
7	4	11

Use a ruler to draw these boxes neatly, and complete the number squares.

1.

3	4	
4	5	
		☐

2.

6	3	
5	5	
		☐

3.

7	2	
3	6	
		☐

4.

4	3	
6	5	
		☐

5.

6	5	
4	7	
		☐

6.

8	3	
1	8	
		☐

7.

7	5	
3	4	
		☐

8.

6	2	
6	7	
		☐

9.

8	4	
6	7	
		☐

10.

2	9	
8	4	
		☐

11.

8	5	
7	6	
		☐

12.

9	6	
5	8	
		☐

13.

8	3	
4	11	
		☐

14.

5	6	
12	5	
		☐

15.

8	6	
11	7	
		☐

16

13	6	
4	9	
		☐

17.

14	6	
3	13	
		☐

18.

7	15	
5	5	
		☐

19.

12	12	
6	9	
		☐

20.

4	16	
8	6	
		☐

Symmetry

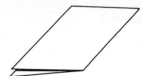 

Fold a piece of paper. . . . Then draw a shape and cut it out.

Unfold the shape. You will have a symmetrical shape. Both halves will be exactly the same.

The crease down the middle of the shape is called the line of symmetry.

Here are some examples. Try to make some of your own.

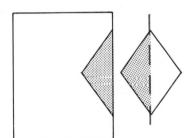 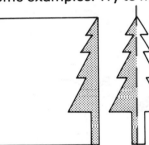 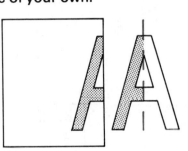

Find the other half of the monster's face so that his face is symmetrical.

A   B   C   D   E

Be careful. All the faces look similar.
If you want to see the monster's face, place a mirror along the broken line of symmetry.

Exercise 1

Here are some pictures. Only one half of each picture is shown. The broken line of symmetry is drawn for you.

Copy each picture and complete it.

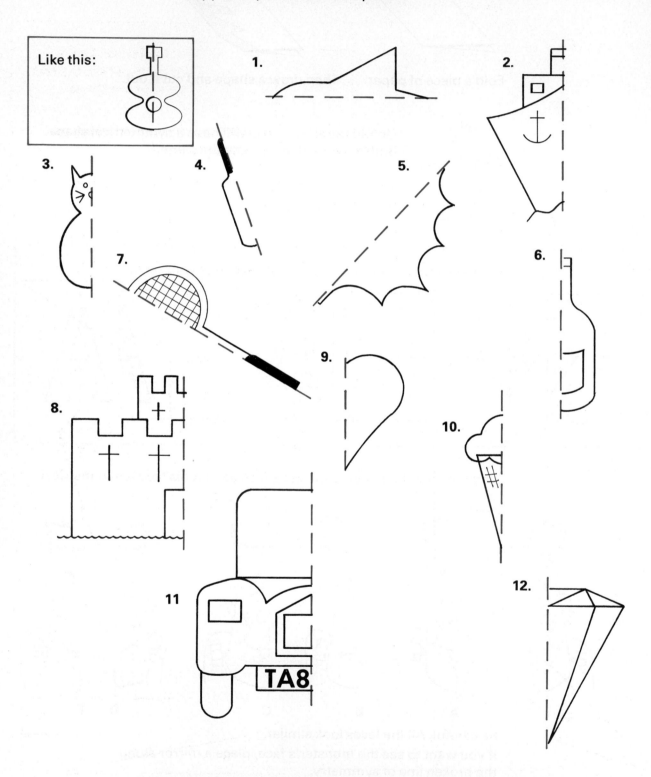

Like this:

Exercise 2

Draw one of the shapes on the left on squared paper.
Choose the matching shape on the right and complete
the shape you have drawn.
Now do the same for all the shapes.

Like this:

A | Y

Put the broken line of
symmetry where the two
halves of the shape
join.

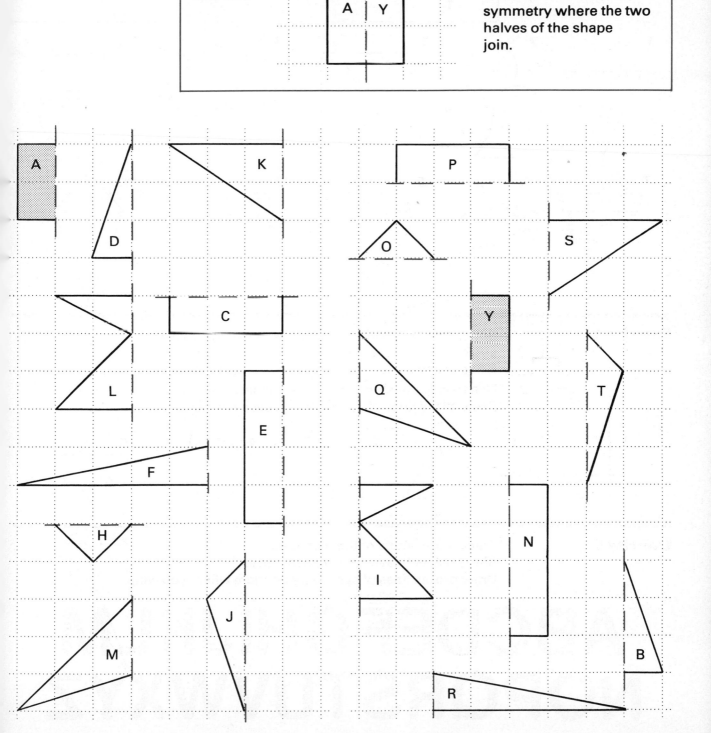

Exercise 3   Which lines of symmetry have been drawn in the correct position?

Answer the questions like this.

 Line 1 is in the
correct position
1.

2. Line 2 is in the
wrong position

1.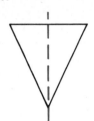

2.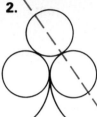

3.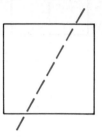

4.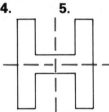

5.

6.

7.

8.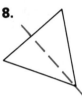

9.

10.

11.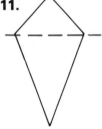

Exercise 4   Some shapes have more than one line of symmetry.
Look at the H shape above.
It has two lines of symmetry.
Use tracing paper to copy these shapes.
Draw all the lines of symmetry on each shape.

1.

2.

3.

4.

5.

Exercise 5   Trace these letters of the alphabet.
Many of them have lines of symmetry.
Draw broken lines of symmetry on them where you can.

ABCDEFGHIJKLM
NOPQRSTUVWXYZ

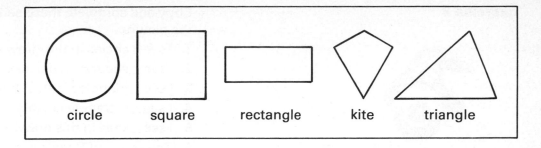

| circle | square | rectangle | kite | triangle |

Exercise 6

Trace the shapes below.
Write the name of each shape by your drawing.

1.  2.   3.   4. 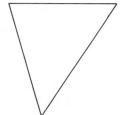 5.

6.   7.  8. 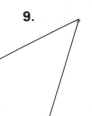 9. 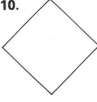 10.

Exercise 7

Copy the table below and tick the shapes that you see in each drawing.
The first one has been done for you.

Drawing	Kite	Triangle	Square	Rectangle	Circle
1.		✔		✔	✔
2.					
3.					
4.					
5.					

1. 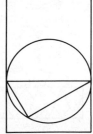 2. 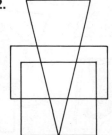 3.  4. 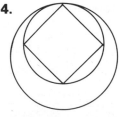 5.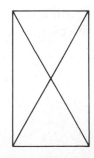

Exercise 8

Copy and complete these sentences about the picture.

1. I see 6 circles in this drawing.
2. I see__squares in this drawing.
3. I see__triangles in this drawing.
4. I see__rectangles in this drawing.
5. I see__kites in this drawing.
6. The shape of Dozo's nose is a _____ .
7. In one hand Dozo is holding three different shapes. These are _____ , _____ and _____ .
8. The shape that Dozo's foot stands on is a _____ .

Exercise 9

Copy and complete these sentences about the picture.

1. Shape A is a _____ .
2. Shapes B and F are _____ .
3. Shape C is a _____ .
4. Shape D is a _____ .
5. Shape E is a _____ .
6. There are _____ kites in the picture.
7. There are _____ circles in the picture.
8. There are _____ rectangles in the picture.

Circles

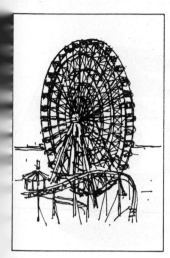

This Big Wheel is the largest in the world. It is $63\frac{1}{2}$ metres (204 feet) high and stands in a fun-fair in Japan.

The radius of the wheel (distance from the centre of the wheel to the edge) is 31 metres (101 feet).

You can draw circles by drawing around the edge of a coin or plate. You can also use a pair of compasses to draw circles carefully.

The distance from the compass point to the tip of the pencil will be the radius of the circle.
What would be the radius of the circle drawn with this pair of compasses?

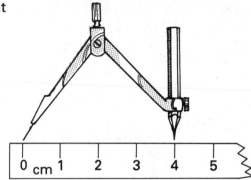

Exercise 10

Measure the radius of each circle.

1.

2.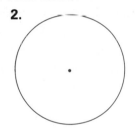

3.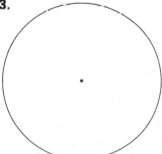

Exercise 11

Draw these circles.

1. Radius 2 cm.
2. Radius 4 cm.
3. Radius 3 cm.
4. Radius 5 cm.
5. Radius 6 cm.
6. Radius $2\frac{1}{2}$ cm.
7. Radius $3\frac{1}{2}$ cm.
8. Radius $1\frac{1}{2}$ cm.
9. Radius $5\frac{1}{2}$ cm.
10. Radius $4\frac{1}{2}$ cm.

Circle patterns

Exercise 12

Here is a pattern made from circles and straight lines.

The drawings below will help you to do your own.

1. In the middle of your page mark a centre point C.
2. With your compass point kept on the point C, draw these circles:
   a. radius 1 cm **b.** radius 2 cm
   c. radius 3 cm **d.** radius 4 cm
   (These are called concentric circles because they have the same centre.)

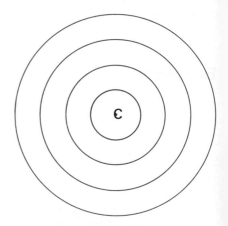

3. Draw straight lines across the largest circle. The lines must be the same distance apart at both ends. Draw the lines so that they just touch the edges of the smaller circles.
4. Think of your own colour scheme to decorate your pattern.

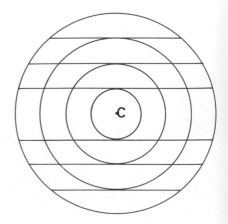

Review 1

A. Measurement

1. Which man is the tallest?
Use a ruler to check your answer.

2. Are these boxes all the same length, or is one longer than the other two?

a.  b.  c.

3. Which line is the longer, 'a' or 'b'?
Use your ruler to check your answer.

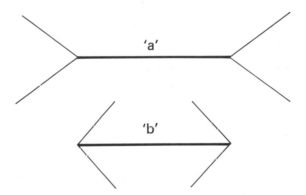

4. Which circle is the bigger, 'a' or 'b'?
Check your answer with a ruler.

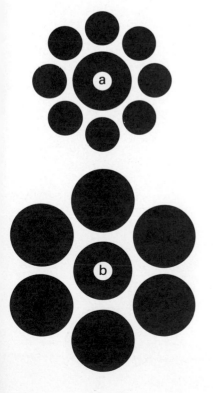

5. Are these two lines straight or crooked?
Check by measuring the distance between them at the top, the bottom and the middle.

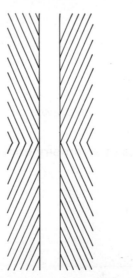

B. Shape

1. This shape is a _____ .

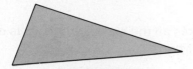

2. This shape is a _____ .

3. This shape is a _____ .

4. This shape is a _____ .

5. What is this shape called?

6. What is this shape called?

7. a. Which line is a radius, a, b, c or d?
b. How long is the radius?

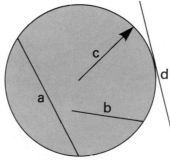

8. Using your compass, ruler and pencil draw a circle of radius 3 cm.

9. Now draw a circle of radius $4\frac{1}{2}$ cm.

10. Copy these shapes. Draw lines of symmetry on them.

a.

b.

c.

d.

This is the Morris Family Tree

Mr. Morris
60 years old

Mrs. Morris
55 years old

Sara
17 years old

Christine
21 years old

Diana
24 years old

Patricia
27 years old

Richard
30 years old

Michael
35 years old

Emily
4 years old

Claire
2 years old

Ricky
1 year old

Helen
3 years old

Exercise 1

Copy and complete the sentences.

1. Helen is _____ years old.
2. Sara is _____ years old.
3. Mr. Morris is _____ years old.
4. Ricky is _____ year old.
5. Mr. and Mrs. Morris have _____ children.
6. Mr. and Mrs. Morris have _____ grandchildren.
7. Altogether there are _____ people in the family.
8. There are _____ males in the family.
9. There are _____ females in the family.
10. There are _____ more females in the family than males.

Exercise 2

Answer these questions.

1. Who is the mother of Emily and Claire?
2. Who is Helen's father?
3. How many children has Richard?
4. How much older is Mr. Morris than his wife?
5. Who is the younger, Emily or Ricky?
6. What is the name of Richard's brother?
7. Who is Claire's sister?
8. Who is Ricky's only uncle?
9. Who is Emily's grandmother?
10. Who is the oldest of Mr. Morris' daughters?

Pat　　Don　　Adam　　Paul　　Linda　　Roy　　Janet

Exercise 3

Look at the picture and answer these questions.

1. Who is the tallest person?
2. Which people are the same height?
3. Who is taller, Pat or Linda?
4. Is Don taller than Linda?
5. Who is taller than Roy?
6. What is the name of the shortest girl?
7. How many people are taller than Pat?
8. Who is shorter than Adam?
9. Who is the shortest person?
10. Who is taller than Pat but shorter than Adam?

Exercise 4

Copy and complete the sentences below using one of the expressions.
The first one has been done for you.

'weighs more than'　　　'weighs less than'　　　'weighs the same as'

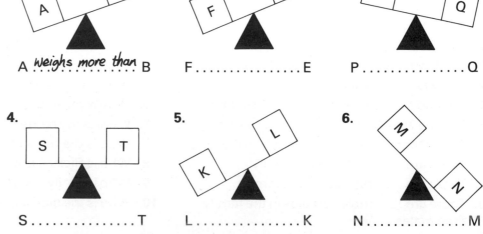

1.

A *weighs more than* B

2.

F . . . . E

3.

P . . . . Q

4.

S . . . . T

5.

L . . . . K

6.

N . . . . M

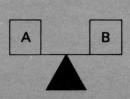 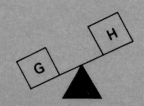

A weighs the same as B
So we can say A = B
This means A is the
same as B.

G weighs more than H
So we say G ≠ H
This means G is not the
same as H.

Exercise 5

Use this sign = instead of 'the same as' and use this sign ≠ instead of 'not the same as' in the questions below.

Like this:
  2 * 1 + 1   4 * 2 + 3
  2 = 1 + 1   4 ≠ 2 + 3

1. 6 * 3 + 3   **2.** 7 * 3 + 4   **3.** 7 * 3 + 3
4. 1 * 1 + 0   **5.** 9 * 2 + 3 + 4   **6.** 8 * 1 + 6 + 1
7. 9 * 8 − 1   **8.** 10 − 3 * 8   **9.** 17 * 5 + 12
10. 20 − 11 * 9   **11.** 9 + 6 + 5 * 22 **12.** 23 − 6 * 18
13. 16 + 5 + 3 * 31 **14.** 42 − 12 * 30   **15.** 27 + 3 − 15 * 15

Exercise 6

Use the = sign and the ≠ again in the questions below.

1. 2 + 1 * 4 − 1   **2.** 5 + 3 * 3 + 1
3. 7 + 3 * 5 + 5   **4.** 10 + 2 * 11 + 3
5. 15 + 1 * 17 − 3   **6.** 2 + 8 + 6 * 8 + 8
7. 29 − 2 * 8 + 8 + 8   **8.** 9 + 2 + 10 * 11 + 12
9. 17 − 17 * 12 − 3 − 9 **10.** 18 + 3 * 13 + 8 + 1
11. 5 + 4 * 13 − 4   **12.** 27 − 3 − 2 * 11 + 11
13. 19 − 18 * 26 − 25   **14.** 9 + 15 * 30 − 6
15. 14 + 17 * 6 + 21   **16.** 28 − 9 * 6 + 9 + 4
17. 15 + 15 * 15 − 15   **18.** 22 + 0 * 22 − 0
19. 6 + 22 * 30 − 2   **20.** 7 − 3 * 49 − 6 − 20

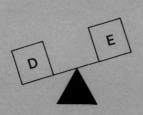
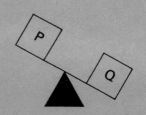

D weighs more than E
So we can say, D > E
This means D is bigger than E

P weighs less than Q
So we say, P < Q
This means P is less than Q

Exercise 7

Use this sign > to mean more than, or use this sign < to mean less than in the questions below.

Like this:
5 * 4   2 * 5
5 > 4   2 < 5

1. 6 * 4   **2.** 10 * 2   **3.** 13 * 11

4. 5 * 11   **5.** 21 * 15   **6.** 16 * 25

7. 33 * 60   **8.** 64 * 72   **9.** 87 * 17

10. 71 * 68   **11.** 101 * 99   **12.** 201 * 199

13. 167 * 200   **14.** 267 * 266   **15.** 461 * 497

Exercise 8

Use the > sign or the < sign in the questions below.

1. 6 + 2 * 9   **2.** 7 + 2 * 8

3. 7 + 3 * 11   **4.** 5 * 6 − 2

5. 6 + 4 * 7   **6.** 7 + 4 * 12

7. 13 * 6 − 1   **8.** 15 * 12 + 5

9. 16 * 20 − 3   **10.** 16 − 3 * 12 − 1

11. 17 + 4 * 13 + 5   **12.** 19 + 2 * 18 + 4

13. 20 + 4 * 16 + 6   **14.** 21 − 2 * 17 + 1

15. 22 + 6 * 5 + 24   **16.** 18 + 9 * 20 + 5

17. 31 − 2 * 24 + 3   **18.** 35 + 1 * 38 − 3

19. 40 − 7 * 31 + 5   **20.** 43 − 10 * 29 + 5

Section 7 — Number patterns

This is Leaping Larry.
Larry leaps along the number line.

What is the size of each jump?   The size of each jump is 2.
Where will the next jump land?   The next jump will land on 11.

Exercise 1   Copy and complete the drawings and the sentences.
The first one is done for you.

1.

The size of each jump is 1.
The next jump will land on 4.

2.

The size of each jump is . . .
The next jump will land on . . .

3.

The size of each jump is . . .
The next jump will land on . . .

4.

The size of each jump is . . .
The next jump will land on . . .

5.

The size of each jump is . . .
The next jump will land on . . .

6.

The size of each jump is . . .
The next jump will land on . . .

7.

The size of each jump is . . .
The next jump will land on . . .

8.

The size of each jump is . . .
The next jump will land on . . .

9.

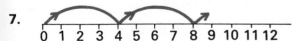

The size of each jump is . . .
The next jump will land on . . .

10.

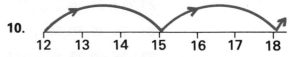

The size of each jump is . . .
The next jump will land on . . .

Larry leaps from 2 on to 6 and then on to 10.

How many spaces did Larry leap with each jump?
He leaped 4 spaces.

Where will Larry land with the next jump?
He will land on 14.

Exercise 2

Here are Larry's jumps. Find out the size of each jump and fill in the missing numbers in your book.

1. 2, 4, 6, 8, ____
2. 3, 5, 7, 9, ____
3. 7, 9, 11, 13, ____
4. 11, 13, 15, 17, ____
5. 0, 3, 6, 9, ____
6. 1, 4, 7, 10, ____
7. 2, 5, 8, 11, ____
8. 4, 7, 10, 13, ____
9. 9, 12, 15, 18, ____
10. 6, 11, 16, 21, ____
11. 10, 16, 22, 28, ____
12. 4, 8, 12, 16, ____
13. 7, 11, 15, 19, ____
14. 13, 16, 19, 22, ____
15. 10, 14, 18, 22, ____
16. 5, 8, 11, 14, ____
17. 5, 9, 13, ____, 21
18. 2, ____, 6, 8, 10
19. 0, ____, 14, 21, 28
20. 10, 20, ____, 40, 50
21. 1, ____, 7, 10, 13, 16
22. 12, ____, 36, 48, 60
23. 12, 24, 36, ____, 60
24. 4, 8, ____, 16, 20
25. 20, 40, 60, 80, ____
26. 9, 18, 27, 36, ____
27. 25, 50, 75, ____, 125
28. 5, 10, 15, ____, ____
29. 30, 50, ____, ____, 110
30. 4, 8, 13, 19, 26, ____

Here Larry is jumping back along the number line.
He has eaten the carrot!

```
2   3   4   5   6   7   8   9   10  11
```

What is the size of each jump?
Where will the next jump land?

The size of each jump is 2.
The next jump will land on 3.

Exercise 3

Copy and complete the drawings and the sentences.
The first one is done for you.

1.

```
0   1   2   3   4   5   6   7
```

The size of each jump is 1.
The next jump will land on 2.

2.

```
8   9   10  11  12  13
```

The size of each jump is . . .
The next jump will land on . . .

3.

```
9   10  11  12  13  14  15  16
```

The size of each jump is . . .
The next jump will land on . . .

4.

```
5       7       9       11
```

The size of each jump is . . .
The next jump will land on . . .

5.

```
12  13  14  15  16  17  18  19  20  21
```

The size of each jump is . . .
The next jump will land on . . .

6.

```
9           12          15          18
```

The size of each jump is . . .
The next jump will land on . . .

7.

```
12  13  14  15  16  17  18  19  20
```

The size of each jump is . . .
The next jump will land on . . .

8.

```
10          14          18
```

The size of each jump is . . .
The next jump will land on . . .

9.

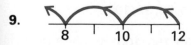

```
8       10      12
```

The size of each jump is . . .
The next jump will land on . . .

10.

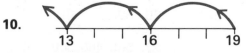

```
13          16          19
```

The size of each jump is . . .
The next jump will land on . . .

Larry is going back again. He starts at 22 and leaps back to 17 and then back to 12.

How many spaces did Larry leap with each jump?   He leaped 5 spaces.
Where will Larry land with the next jump?   He will land on 7.

Exercise 4   Here are Larry's jumps. Find out the size of each jump and fill in the missing numbers in your book.

1. 10, 8, 6, 4, ____   2. 15, 12, 9, 6, ____

3. 22, 20, 18, 16, ____   4. 9, 7, 5, 3, ____

5. 27, 25, 23, 21, ____   6. 13, 11, 9, 7, ____

7. 21, 18, 15, 12, ____   8. 17, 15, 13, 11, ____

9. 24, 20, 16, 12, ____   10. 40, 30, 20, 10, ____

11. 25, 20, 15, 10, ____   12. 17, 14, 11, 8, ____

13. 70, 50, 30, ____   14. 20, 16, 12, 8, ____

15. 26, 22, 18, 14, ____   16. 40, ____, 20, 10, 0

17. 55, ____, 45, 40, 35   18. 80, ____, 40, 20, 0

19. 28, 21, 14, ____, 0   20. 18, 15, 12, ____, 6

21. 21, 17, 13, ____, 5   22. 11, 9, ____, 5, 3

23. 26, 20, 14, ____, 2   24. 29, 26, ____, 20, 17

25. 60, ____, 40, 30, 20   26. ____, 70, 50, 30, 10

27. 500, 400, 300, 200, ____   28. 250, 200, 150, 100, ____

29. 150, 120, 90, 60, ____   30. 110, 80, ____, 20,

Addition

```
                    Workcard 1

1.    26        2.    30        3.    40
     +40             +50             +30

4.    35        5.    17        6.    30
     +15             +23             +60

7.    21        8.    10        9.    62
     +39             +50             +28

10.   14       11.    33       12.    51
     +26             +27             +19
```

```
                    Workcard 2

1.     5        2.    72        3.    65
     +27             +90             +19

4.    37        5.    38        6.    18
     +36             +4              +43

7.    49        8.     7        9.    55
     +6              +25             +8

10.    8       11.    19       12.    74
     +54             +41             +6
```

Exercise 1

```
1.    12      2.    23      3.    34      4.    26      5.    14      6.    30
      21            10            14             2            20            11
     +14            +6           +21           +11           +15           +27


7.    25      8.    16      9.    24     10.     8     11.    26     12.    13
       5            12            16            12            16            18
     +20            +8            +7           +25           +25           +36
```

13. In a car park, there are 5 red cars, 2 blue cars and 2 yellow cars. How many cars are there in the car park?
    Answer: There are _____ cars in the car park.

14. Mary found 5 pencils in one box, 6 pencils in another and 2 pencils in another. How many pencils has Mary found?
    Answer: Mary has found _____ pencils.

15. Jenny has 8 pink ribbons, 6 blue ribbons and 5 green ribbons. How many ribbons has she altogether?
    Answer: Jenny has _____ ribbons in all.

16. Robin spent 23p on a comic and 9p on a sweet. How much did he spend?
    Answer: Robin spent _____ p.

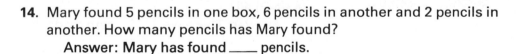

17. With three darts, Tony scored 9, 15 and 12. What was his total score?
    Answer: Tony scored _____ in all.

18. In three days, Bob saved 26p, 14p and 28p. How much did he save in all?
    Answer: Bob saved _____ p in all.

19. In her last three spelling tests, Mary scored 7, 10 and 9. What is her total score?

    Answer: Mary scored _____ marks altogether.

20. Ali has 15 marbles, Tracy has 13 marbles and John has 8. How many marbles have they in all?

    Answer: They have _____ marbles in all.

21. David was given 9p by his mother, 16p by his uncle and 10p by his aunt. How much was David given in total?

    Answer: David was given _____ p.

22. Rosemary baked 22 jam tarts, 14 apple tarts and 16 custard tarts. How many tarts did she bake?

    Answer: Rosemary baked _____ tarts.

23. There are 18 pupils in one classroom and 6 more walk in. Later 12 more walk in. How many pupils are there in the classroom?

    Answer: There are _____ pupils in the classroom.

24. Roy has three bags of sweets. In the first bag he has 14 sweets, in the second bag he has 23 sweets and in the third he has 17. How many sweets has Roy?

    Answer: Roy has _____ sweets in all.

25. In a box there are 14 red crayons, 26 green ones, 5 blue ones and 10 yellow ones. How many crayons are there altogether?

    Answer: There are _____ crayons in the box.

26. Ian has 17 comics. Tommy has 16 comics and James has 5. How many comics have they in all?

    Answer: The boys have _____ comics altogether.

27. A motor car factory made 16 cars on Monday, 19 cars on Tuesday, 20 cars on Wednesday and 22 cars on Thursday. How many cars were made in all?

    Answer: There were _____ cars made during the four days.

28. Three coaches arrive at a soccer match: 31 fans were in the first coach, 26 in the second and 25 in the third. How many fans were in the coaches?

    Answer: There were _____ fans in the three coaches.

29. There are 29 pupils in class 'A', 26 pupils in class 'B', 23 pupils in class 'C' and 21 pupils in class 'D'. How many pupils are there in all?

    Answer: There are _____ pupils altogether.

30. In their last game, the players in the school cricket team made these scores: Patel scored 12, Morris scored 5, Bright made 14, May scored 6, Williams scored 20, Allan did not score. What was the total?

    Answer: The cricket team scored _____ runs.

Subtraction

Exercise 2

Solve these problems. See how many you can do in 30 minutes.

1. If you took 2 apples from this tray, how many would be left?

2. If you took 5 sweets from this tray, how many would be left?

3. If you took 4 mushrooms from this tray, how many would be left?

4. If you took 3 boxes from this tray, how many would be left?

Copy out these sums. What number or sign is missing from each box?

5.
$$\begin{array}{r} 5 \\ -3 \\ \hline \Box \end{array}$$

6.
$$\begin{array}{r} 4 \\ -\Box \\ \hline 1 \end{array}$$

7.
$$\begin{array}{r} 6 \\ -2 \\ \hline \Box \end{array}$$

8.
$$\begin{array}{r} 3 \\ -2 \\ \hline \Box \end{array}$$

9.
$$\begin{array}{r} 8 \\ -\Box \\ \hline 6 \end{array}$$

10.
$$\begin{array}{r} 5 \\ \Box 4 \\ \hline 1 \end{array}$$

11.
$$\begin{array}{r} 7 \\ \Box 3 \\ \hline 4 \end{array}$$

12.
$$\begin{array}{r} 9 \\ -\Box \\ \hline 4 \end{array}$$

13.
$$\begin{array}{r} \Box \\ -3 \\ \hline 5 \end{array}$$

14.
$$\begin{array}{r} \Box \\ -4 \\ \hline 4 \end{array}$$

15.
$$\begin{array}{r} 7 \\ -2 \\ \hline \Box \end{array}$$

16.
$$\begin{array}{r} 6 \\ -\Box \\ \hline 3 \end{array}$$

17.
$$\begin{array}{r} 4 \\ \Box 2 \\ \hline 2 \end{array}$$

18.
$$\begin{array}{r} \Box \\ -3 \\ \hline 6 \end{array}$$

19. 6 take away 2 is _____
20. 8 minus 5 is _____
21. 9 take away 1 equals _____
22. 7 minus 6 equals _____
23. 5 take away 5 is _____
24. 10 minus 5 is _____
25. 10 minus 1 equals _____
26. 12 take away 6 equals _____
27. 13 take away 3 is _____
28. 11 take away 5 makes _____
29. 9 take away 7 is _____
30. 12 subtract 10 equals _____

31. There are 15 toffees in this bag.

Ben eats 3 and Judy eats 4; how many are left?

There are _____ toffees left.

32. Steven has 17 comics.

He gives 5 of them to Lenny and 2 to Bob. How many comics has he got left?

Steven has _____ comics left.

33. Mary has 24p in her piggy bank.

She takes 10p out on Monday and 7p out on Tuesday. How much has she left?

Mary has _____ p left in her piggy bank.

34. Mark had 28p, so he went to the sweet shop.

He bought a packet of gum for 8p and a sweet for 10p. How much money has he left?

Mark has _____ p left.

Workcard 1

1. 5
   −2

2. 4
   −3

3. 6
   −4

4. 9
   −5

5. 9
   −1

6. 7
   −6

7. 8
   −2

8. 8
   −5

9. 6
   −3

10. 6
    −1

11. 7
    −4

12. 5
    −4

Workcard 2

1. 8
   −3

2. 9
   −8

3. 7
   −2

4. 6
   −2

5. 3
   −2

6. 5
   −1

7. 8
   −4

8. 6
   −0

9. 7
   −3

10. 8
    −0

11. 9
    −5

12. 7
    −0

Workcard 3

1. 24
   −10

2. 26
   −13

3. 26
   −12

4. 37
   −24

5. 28
   −16

6. 29
   −11

7. 25
   −14

8. 23
   −11

9. 27
   −16

10. 29
    −16

11. 28
    −14

12. 48
    −15

Workcard 4

1. 36
   −21

2. 43
   −30

3. 56
   −12

4. 44
   −33

5. 52
   −31

6. 67
   −23

7. 35
   −15

8. 61
   −30

9. 54
   −24

10. 47
    −15

11. 63
    −43

12. 37
    −20

Workcard 5

1. 15
   −9

2. 14
   −7

3. 16
   −7

4. 12
   −4

5. 13
   −8

6. 15
   −7

7. 17
   −9

8. 12
   −3

9. 11
   −7

10. 11
    −3

11. 22
    −8

12. 13
    −7

Workcard 6

1. 21
   −19

2. 33
   −26

3. 24
   −15

4. 25
   −7

5. 32
   −26

6. 37
   −28

7. 25
   −10

8. 45
   −20

9. 45
   −38

10. 26
    −10

11. 34
    −10

12. 36
    −17

	Workcard 7				
1.	25 −9	**2.**	21 −3	**3.**	25 −6
4.	32 −9	**5.**	24 −7	**6.**	21 −6
7.	36 −7	**8.**	31 −5	**9.**	38 −9
10.	26 −8	**11.**	27 −9	**12.**	23 −5

	Workcard 8				
1.	10 −6	**2.**	20 −6	**3.**	20 −9
4.	40 −21	**5.**	50 −39	**6.**	60 −41
7.	30 −16	**8.**	60 −24	**9.**	50 −35
10.	30 −18	**11.**	80 −27	**12.**	80 −61

Exercise 3   Complete these problems. The first one has been done for you.

1.
$\left(12 \right)$
− 6 = 6
− 4 = 8
− 2 = 10

2.
$\left(15 \right)$
− 5 = *
− 12 = *
− 4 = *

3.
$\left(10 \right)$
− 9 = *
− 4 = *
− 1 = *

4.
$\left(13 \right)$
− 0 = *
− 9 = *
− 13 = *

5.
$\left(11 \right)$
− 2 = *
− 7 = *
− 9 = *

6.
$\left(27 \right)$
− 11 = *
− 24 = *
− 20 = *

7.
$\left(34 \right)$
− 33 = *
− 16 = *
− 24 = *

8.
$\left(30 \right)$
− 17 = *
− 30 = *
− 21 = *

9.
$\left(43 \right)$
− 16 = *
− 24 = *
− 26 = *

Multiplication

Farmer Jones has 4 stalls. In each stall there are 3 pigs.
To find out how many pigs he has, Farmer Jones adds 4 lots of pigs

Like this: 3 + 3 + 3 + 3 = 12 pigs

If he had 5 stalls with three pigs in each stall, he would have
5 lots of pigs.

Like this: 3 + 3 + 3 + 3 + 3 = 15 pigs

Exercise 4

Write out these problems in the same way as Farmer Jones.
Give the answer like this:-
    3 lots of 3 = 3 + 3 + 3 = 9

1. 2 lots of 3	**2.** 2 lots of 5	**3.** 5 lots of 2	**4.** 3 lots of 2
5. 4 lots of 3	**6.** 2 lots of 6	**7.** 3 lots of 5	**8.** 2 lots of 7
9. 4 lots of 4	**10.** 6 lots of 3	**11.** 4 lots of 2	**12.** 6 lots of 2

Exercise 5   Two times

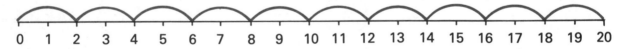

1. What do 3 jumps of 2 make?   **2.** What do 2 jumps of 2 make?   **3.** What do 5 jumps of 2 make?
4. What do 6 jumps of 2 make?   **5.** What do 9 jumps of 2 make?   **6.** What do 8 jumps of 2 make?

7. There are 2 beads in each tin.
How many beads are there in 7 tins?

8. There are 2 people in each car.
How many people are there in 5 cars?

Use the number line to solve these problems.

9. 8, * , 12, * , 16, *   **10.** 10, * , * , 16, *   **11.** 14, * , 18, *
12. a. 1 × 2 = *   **b.** 3 × 2 = *   **c.** 5 × 2 = *   **d.** 4 × 2 = *
   e. 6 × 2 = *   **f.** 8 × 2 = *   **g.** 7 × 2 = *   **h.** 10 × 2 = *

Exercise 6 Three times

1. What do 3 jumps of 3 make?
2. What do 2 jumps of 3 make?
3. What do 6 jumps of 3 make?
4. What do 8 jumps of 3 make?
5. There are 3 sweets in each box. How many sweets in 4 boxes?
6. There are 3 apples in each tin. How many apples in 7 tins?

Use the number line to solve these problems.
7. 6, 9, *, 15, *, 21
8. 12, *, 18, *, 24
9. 15, *, *, 24, *
10. a. $1 \times 3 = *$ b. $3 \times 3 = *$ c. $* \times 3 = 12$ d. $* \times 3 = 27$

Exercise 7 Four times

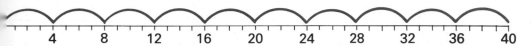

1. What does 1 jump of 4 make?
2. What do 3 jumps of 4 make?
3. What do 4 jumps of 4 make?
4. What do 8 jumps of 4 make?
5. There are 4 cakes in each pack. How many cakes in 3 packs?
6. There are 4 biscuits in each box. How many are there in 6 boxes?

Use the number line to solve these problems.
7. 0, 4, *, 12, *
8. *, 16, *, 24, *
9. 24, *, 32, *, *
10. a. $2 \times 4 = *$ b. $8 \times 4 = *$ c. $* \times 4 = 12$ d. $* \times 4 = 24$

Exercise 8 Five times

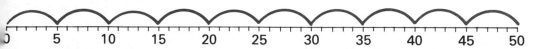

1. What do 3 jumps of 5 make?
2. What do 5 jumps of 5 make?
3. What do 7 jumps of 5 make?
4. What do 10 jumps of 5 make?
5. There are 5 sweets in each pack. How many sweets in 2 packs?
6. There are 5 eggs in each box. How many eggs in 4 boxes?

Use the number line to solve these problems.
7. 5, 10, *, *, 25
8. 20, *, 30, *, 40
9. 35, *, 45, *, *
10. a. $3 \times 5 = *$ b. $8 \times 5 = *$ c. $* \times 5 = 10$ d. $* \times 5 = 35$

Exercise 9 Six times

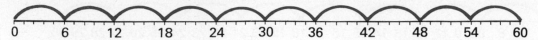

| 0 | 6 | 12 | 18 | 24 | 30 | 36 | 42 | 48 | 54 | 60 |

1. What do 3 jumps of 6 make?
2. What do 5 jumps of 6 make?
3. What do 6 jumps of 6 make?
4. What do 9 jumps of 6 make?
5. There are 6 plums in each tin. How many plums in 3 tins?
6. There are 6 stones in each pot.  How many stones in 7 pots?

Use the number line to solve these problems.

7. 0, 6, * 18, *
8. *, 24, 30, *, 42
9. 36, *, *, 54, *
10. a. $4 \times 6 = *$ b. $8 \times 6 = *$ c. $* \times 6 = 60$ d. $* \times 6 = 36$

Workcard 1		
1. 2	**2.** 3	**3.** 4
3 ×	3 ×	2 ×
4. 6	**5.** 3	**6.** 2
2 ×	5 ×	7 ×
7. 4	**8.** 5	**9.** 8
4 ×	4 ×	2 ×
10. 9	**11.** 6	**12.** 6
2 ×	5 ×	4 ×

Workcard 2		
1. 4	**2.** 8	**3.** 6
7 ×	3 ×	6 ×
4. 5	**5.** 7	**6.** 5
6 ×	5 ×	5 ×
7. 5	**8.** 3	**9.** 9
1 ×	7 ×	4 ×
10. 2	**11.** 3	**12.** 5
10 ×	9 ×	10 ×

Workcard 3		
1. 6	**2.** 4	**3.** 8
3 ×	5 ×	4 ×
4. 1	**5.** 6	**6.** 8
6 ×	7 ×	5 ×
7. 8	**8.** 5	**9.** 6
6 ×	9 ×	10 ×
10. 4	**11.** 7	**12.** 9
10 ×	7 ×	6 ×

Exercise 10

Copy out these targets and fill in the missing numbers.
The first one has been done for you.

1.

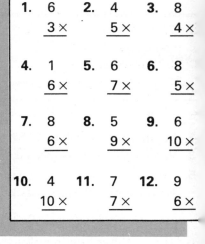

2.

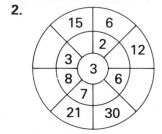

3.

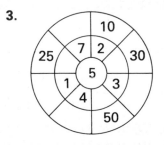

4.

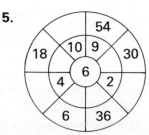

5.

Division

Farmer Jones has 15 hens. He wants to share the hens equally between 3 hen houses.

15 shared between 3 = 5.

Therefore there will be 5 hens in each hen house.

Exercise 11 Copy and complete the following. The first one is done for you.

1. 6 shared between 2 is 3
2. 8 shared between 4 is *
3. 8 shared between 2 is *
4. 12 shared between 3 is *
5. 10 shared between 2 is *
6. 6 shared between 3 is *
7. 10 shared between 5 is *
8. 14 shared between 7 is *
9. 9 shared between 9 is *

Exercise 12

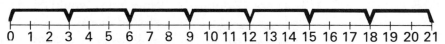

0 1 2 3 4 5 6 7 8 9 10 11 12 13 14 15 16 17 18 19 20 21

1. How many groups of 3 will you have in 9?
2. How many groups of 3 will you have in 15?
3. How many groups of 3 will you have in 6?
4. How many groups of 3 will you have in 21?
5. How many groups of 3 will you have in 18?

Exercise 13

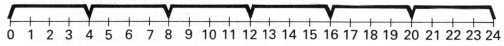

0 1 2 3 4 5 6 7 8 9 10 11 12 13 14 15 16 17 18 19 20 21 22 23 24

1. How many groups of 4 will you have in 8?
2. How many groups of 4 will you have in 16?
3. How many groups of 4 will you have in 12?
4. How many groups of 4 will you have in 24?
5. How many groups of 4 will you have in 20?

Exercise 14

0 5 10 15 20 25 30

1. How many groups of 5 will you have in 15?
2. How many groups of 5 will you have in 20?
3. How many groups of 5 will you have in 10?
4. How many groups of 5 will you have in 30?
5. How many groups of 5 will you have in 25?

Exercise 15

1. Share these 6 apples between 2 baskets.
   Answer: There will be ____ apples in each basket.

2. Share these 9 sweets between 3 children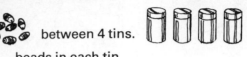
   Answer: Each child will have ____ sweets.

3. Share these 12 beads between 4 tins.
   Answer: There will be ____ beads in each tin.

4. Share these 10 buns between 2 boxes.
   Answer: There will be ____ buns in each box.

5. Share these 16 coins  between 4 children.
   Answer: Each child will get ____ coins.

Exercise 16

1. Here are 12 dots.

   How many would be in each group if you shared them
   a. between 2 groups   **b.** between 4 groups
   c. between 6 groups?

2. Here are 16 crosses.

   How many would be in each group if you shared them
   a. between 2 groups   **b.** between 4 groups
   c. between 8 groups?

3. Here are 18 squares.
   How many would be in each group if you shared them
   a. between 9 groups   **b.** between 3 groups
   c. between 2 groups?

Workcard 1		
1. 2)6̄	**2.** 3)6̄	**3.** 2)8̄
4. 3)9̄	**5.** 2)10̄	**6.** 3)12̄
7. 4)12̄	**8.** 5)10̄	**9.** 2)12̄
10. 3)15̄	**11.** 4)16̄	**12.** 5)15̄
13. 5)20̄	**14.** 6)18̄	**15.** 4)24̄

Workcard 2		
1. $6 \div 2 =$	**2.** $9 \div 3 =$	**3.** $8 \div 2 =$
4. $8 \div 4 =$	**5.** $6 \div 3 =$	**6.** $10 \div 2 =$
7. $10 \div 5 =$	**8.** $12 \div 4 =$	**9.** $12 \div 6 =$
10. $14 \div 2 =$	**11.** $15 \div 5 =$	**12.** $16 \div 4 =$
13. $20 \div 5 =$	**14.** $18 \div 6 =$	**15.** $20 \div 10 =$

| + means add | − means take away | × means times | ÷ means divide |

Exercise 17   Copy out these problems and put the right sign in place of ∗

1. 3 ∗ 2 = 5   **2.** 6 ∗ 2 = 4   **3.** 3 ∗ 2 = 6   **4.** 8 ∗ 3 = 5

5. 8 ∗ 2 = 4   **6.** 3 ∗ 3 = 9   **7.** 8 ∗ 4 = 12   **8.** 6 ∗ 2 = 3

9. 9 ∗ 1 = 8   **10.** 5 ∗ 2 = 10   **11.** 10 ∗ 2 = 5   **12.** 6 ∗ 5 = 11

13. 8 ∗ 2 = 6   **14.** 3 ∗ 4 = 12   **15.** 8 ∗ 9 = 17   **16.** 8 ∗ 2 = 16

17. 8 ∗ 4 = 2   **18.** 4 ∗ 4 = 16   **19.** 9 ∗ 3 = 6   **20.** 4 ∗ 16 = 20

21. 6 ∗ 8 = 14   **22.** 5 ∗ 3 = 15   **23.** 9 ∗ 3 − 3   **24.** 9 × 9 = 0

Exercise 18   Can you work out the message below from this code table?

A	B	C	D	E	F	G	H	I	J	K	L	M
24	39	21	8	13	14	1	18	16	23	4	11	20

N	O	P	Q	R	S	T	U	V	W	X	Y	Z
30	2	6	5	25	9	7	15	19	3	12	17	28

(4 + 3 = ∗)   (6 × 3 = ∗)   (19 − 6 = ∗)

(10 ÷ 2 = ∗)   (5 × 3 = ∗)   (27 − 11 = ∗)   (12 + 9 = ∗)   (12 ÷ 3 = ∗)

(8 + 31 = ∗)   (5 × 5 = ∗)   (17 − 15 = ∗)   (6 ÷ 2 = ∗)   (10 × 3 − ∗)

(21 − 7 = ∗)   (20 ÷ 10 = ∗)   (6 + 6 = ∗)

(12 + 11 = ∗)   (20 − 5 = ∗)   (4 × 5 = ∗)   (12 ÷ 2 = ∗)   (3 × 3 = ∗)

(2 × 1 = ∗)   (9 + 10 = ∗)   (26 ÷ 2 = ∗)   (9 + 16 = ∗)

(14 ÷ 2 = ∗)   (9 + 9 = ∗)   (13 + 0 = ∗)

(17 − 6 = ∗)   (8 × 3 = ∗)   (4 × 7 = ∗)   (4 + 13 = ∗)

(16 ÷ 2 = ∗)   (31 − 29 = ∗)   (17 − 16 = ∗)

Exercise 19   Try to put this sentence into code like that in the last
exercise.

The noisy noise annoys an oyster.

Coins

1p
Penny

2p
Two pence

5p
Five pence

10p
Ten pence

20p
Twenty pence

50p
Fifty pence

£1
One pound coin

Notes

£5
Five pound note

£10
Ten pound note

£20
Twenty
pound note

£50
Fifty
pound
note

Exercise 1

How much money is there in each box?

1. 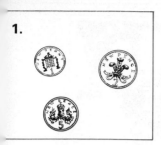	**2.** 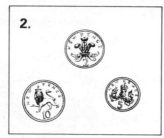	**3.** 	**4.** 
5. 	**6.** 	**7.** 	**8.** 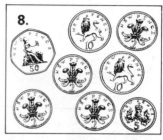
9. 	**10.** 	**11.** 	**12.** 

Exercise 2

1. If you had 4 , how much would you have in all?

2. If you had 5 , how much would you have in all?

3. If you had 3 , how much would you have in all?

4. If you had 7 , how much would you have in all?

5. If you had 4 , how much would you have in all?

6. If you had 12 , how much would you have in all?

7. If you had 8 , how much would you have in all?

8. If you had 4 , how much would you have in all?

9. If you had 25 , how much would you have in all?

10. If you had 3 and 3 , how much would you have in all?

How much money is in each box?

1.

2.

3.

4.

5.

6.

7.

8.

Exercise 4

What is the cost of the following?

crisps 12p
GUM 7p
LEMONADE 26p
Chocolate 23p
CHEWS 9p
Toffee 18p

1. Two packets of gum.
2. A bag of crisps and a pack of gum.
3. A pack of chews and a pack of gum.
4. A bag of crisps and a bar of chocolate.
5. A bar of chocolate and a pack of chews.
6. Two cans of lemonade and a bag of crisps.
7. A bar of chocolate and a can of lemonade.
8. A bar of toffee and a pack of gum.
9. Two bars of chocolate and a pack of chews.
10. A bar of toffee and a can of lemonade.
11. Two packs of gum and two bags of crisps.
12. Two cans of lemonade and a bar of toffee.
13. Two packs of chews, a bar of toffee and a can of lemonade.
14. Two bags of crisps, two packs of gum and a bar of toffee.
15. A bar of chocolate, a bar of toffee and three packs of gum.

Exercise 5

Add up these bills.

1.	2.	3.	4.
A pack of chews . . . . A pack of gum . . . . A bag of crisps . . . . Total:_____	A bar of chocolate . . . . A bag of crisps . . . . A pack of gum . . . . Total:_____	A bar of toffee . . . . A pack of chews . . . . A bar of chocolate . . . . Total:_____	A can of lemonade . . . . A pack of chews . . . . A pack of gum . . . . Total:_____
5.	**6.**	**7.**	**8.**
Two bags of crisps . . . . A pack of gum . . . . A bar of chocolate . . . . Total:_____	A can of lemonade . . . . Two bags of crisps . . . . Two packs of gum . . . . Total:_____	A bar of chocolate . . . . A pack of toffee . . . . Two packs of chews . . . . Total:_____	Two packs of chews . . . . Two packs of gum . . . . A can of lemonade . . . . Total:_____

Exercise 6

CLOCK
£7.00

VASE
£2.50

LAMP
£8.00

KETTLE
£3.50

UMBRELLA
£2.25

1. How much would a kettle and a lamp cost?
2. How much would a clock and a vase cost?
3. How much would a clock and a lamp cost?
4. How much would a vase and a lamp cost?
5. How much would two vases and a lamp cost?
6. How much would an umbrella and a clock cost?
7. How much would two kettles and a clock cost?
8. How much would a vase and an umbrella cost?
9. How much would a vase, a lamp and a kettle cost?
10. How much would two vases and an umbrella cost?
11. How much change from £10 would you get if you bought a lamp?
12. How much change from £10 would you get if you bought a clock?

Exercise 7

Write out these sentences and complete them.

1. There are _____10p pieces in £1.00.
2. There are _____2p pieces in 10p.
3. There are _____5p pieces in 20p.
4. There are _____5p pieces in 50p.
5. There are _____50p pieces in £2.00.
6. There are _____5p pieces in £1.00.
7. There are _____2p pieces in 20p.
8. There are _____20p pieces in £1.00.
9. There are _____20p pieces in £1.60.
10. There are _____10p pieces in £1.40.
11. There are _____5p pieces in 45p.
12. There are _____2p pieces in 34p.

Earning pocket money

To earn their pocket money Jenny and Mark have to help with the housework.
In the kitchen is a chart which says how much they can earn for each task.

Helping with the washing up	25p
Running a message	10p
Taking the dog for a walk	15p
Helping with the ironing	30p
Helping wash the car	30p
Helping in the garden	25p
Tidying up their bedrooms	12p

Exercise 8

Their mother keeps a record of the tasks Jenny and Mark do during each week. Each time a task is done she puts a tick on the chart. Here is the chart for one week.

Task	Jenny	Mark
Washing up	✓ ✓ ✓	✓ ✓
Messages	✓ ✓ ✓	✓ ✓ ✓
Walking dog	✓ ✓	✓ ✓ ✓
Ironing	✓ ✓	✓
Washing car	✓	✓
Gardening	✓	✓
Tidying bedroom	✓ ✓	✓ ✓ ✓

Copy and complete these sentences.
1. Jenny earned _____ p washing up.
2. Mark earned _____ p running messages.
3. Jenny earned _____ p walking the dog.
4. Mark earned _____ p tidying his bedroom.
5. Jenny earned 60p from _____ .
6. Mark earned 45p from _____ .
7. Jenny earned £ _____ in all.
8. Mark earned £ _____ in all.

Exercise 9

Each week Jenny and Mark save 40p.
Copy and complete their savings record chart.
Jenny started with savings of £1.20 and Mark started with 85p.

	Week One	Week Two	Week Three	Week Four	Week Five	Week Six	Week Seven	Week Eight	Week Nine	Week Ten
Jenny	£1.20	*	£2.00	*	*	*	£3.60	*	*	£4.80
Mark	85p	£1.25	*	*	£2.45	*	*	*	£4.05	*

Workcard 1

1.
£	:	p
1		25
+ 2		23

2.
£	:	p
2		14
+ 1		24

3.
£	:	p
3		16
+ 5		42

4.
£	:	p
2		31
+ 6		27

5.
£	:	p
2		45
+ 1		23

6.
£	:	p
4		17
+ 3		61

7.
£	:	p
5		62
+ 2		23

8.
£	:	p
4		20
+ 4		49

Workcard 2

1.
£	:	p
12		41
+15		22

2.
£	:	p
23		37
+42		42

3.
£	:	p
21		33
+14		25

4.
£	:	p
16		51
+23		47

5.
£	:	p
22		41
+ 7		26

6.
£	:	p
6		30
+20		27

7.
£	:	p
41		21
+20		05

8.
£	:	p
28		30
+11		07

Workcard 3

1.
£	:	p
5		20
+ 5		13

2.
£	:	p
15		41
+ 6		07

3.
£	:	p
14		28
+ 7		11

4.
£	:	p
6		34
+ 7		23

5.
£	:	p
9		22
+23		07

6.
£	:	p
16		54
+ 8		20

7.
£	:	p
41		61
+ 9		15

8.
£	:	p
19		07
+28		32

Workcard 4

1.
£	:	p
5		06
+ 3		07

2.
£	:	p
2		16
+ 5		28

3.
£	:	p
14		08
+ 6		23

4.
£	:	p
5		16
+13		25

5.
£	:	p
22		18
+13		25

6.
£	:	p
35		29
+12		17

7.
£	:	p
5		17
+13		46

8.
£	:	p
22		52
+ 6		18

Workcard 5

1.
£	:	p
3		52
+ 1		65

2.
£	:	p
2		61
+ 4		62

3.
£	:	p
1		70
+ 4		41

4.
£	:	p
5		92
+ 2		23

5.
£	:	p
2		72
+ 4		50

6.
£	:	p
5		66
+ 1		71

7.
£	:	p
2		77
+ 0		81

8.
£	:	p
5		60
+ 3		84

Workcard 6

1.
£	:	p
12		55
+ 4		60

2.
£	:	p
3		62
+21		77

3.
£	:	p
4		68
+ 1		73

4.
£	:	p
3		85
+ 5		47

5.
£	:	p
2		86
+ 4		44

6.
£	:	p
5		92
+ 2		38

7.
£	:	p
9		58
+ 2		66

8.
£	:	p
14		83
+ 6		38

The four seasons

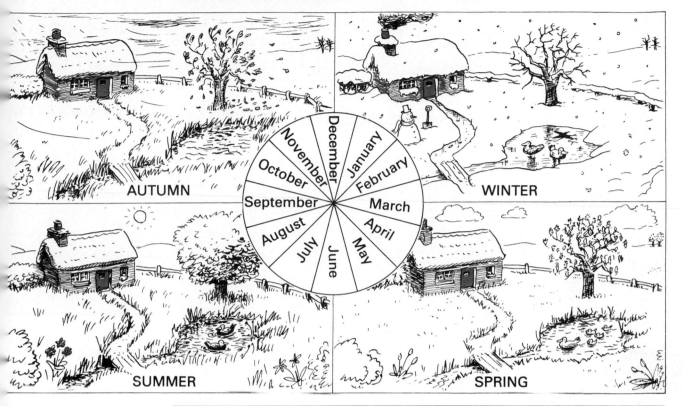

AUTUMN   WINTER   SUMMER   SPRING

The year is divided into four seasons: Spring, Summer, Autumn, Winter. The seasons do not change at the beginning of months, but in the middle of certain months.

Exercise 1   Using the drawing above, answer these questions:

1. In which month do the seasons change from Summer to Autumn?
2. In which month do the seasons change from Winter to Spring?
3. In which month do the seasons change from Autumn to Winter?
4. In which month do the seasons change from Spring to Summer?

Exercise 2   Copy out the statements below. Beside each statement write the name of the season that seems most suitable.

1. Very long nights and short days.
2. Sun tan oil.
3. The season after Summer.
4. Flowers and trees start growing again.
5. Ice and snow.
6. The season that follows Winter.
7. The hottest season.
8. The season in which Christmas falls.
9. Trees lose their leaves and it gets colder.
10. The season between Summer and Winter.

Using a calendar

	January	February	March	April
Sunday	. 3 10 17 24 31	. 7 14 21 28	. 7 14 21 28	. 4 11 18 25
Monday	. 4 11 18 25 .	1 8 15 22 .	1 8 15 22 29	. 5 12 19 26
Tuesday	. 5 12 19 26 .	2 9 16 23 .	2 9 16 23 30	. 6 13 20 27
Wednesday	. 6 13 20 27 .	3 10 17 24 .	3 10 17 24 31	. 7 14 21 28
Thursday	. 7 14 21 28 .	4 11 18 25 .	4 11 18 25 .	1 8 15 22 29
Friday	1 8 15 22 29 .	5 12 19 26 .	5 12 19 26 .	2 9 16 23 30
Saturday	2 9 16 23 30 .	6 13 20 27 .	6 13 20 27 .	3 10 17 24 .

	May	June	July	August
Sunday	. 2 9 16 23 30	. 6 13 20 27	. 4 11 18 25	1 8 15 22 29
Monday	. 3 10 17 24 31	. 7 14 21 28	. 5 12 19 26	2 9 16 23 30
Tuesday	. 4 11 18 25 .	1 8 15 22 29	. 6 13 20 27	3 10 17 24 31
Wednesday	. 5 12 19 26 .	2 9 16 23 30	. 7 14 21 28	4 11 18 25 .
Thursday	. 6 13 20 27 .	3 10 17 24 .	1 8 15 22 29	5 12 19 26 .
Friday	. 7 14 21 28 .	4 11 18 25 .	2 9 16 23 30	6 13 20 27 .
Saturday	1 8 15 22 29 .	5 12 19 26 .	3 10 17 24 31	7 14 21 28 .

	September	October	November	December
Sunday	. 5 12 19 26	. 3 10 17 24 31	. 7 14 21 28	. 5 12 19 26
Monday	. 6 13 20 27	. 4 11 18 25 .	1 8 13 22 29	. 6 13 20 27
Tuesday	. 7 14 21 28	. 5 12 19 26 .	2 9 16 23 30	. 7 14 21 28
Wednesday	1 8 15 22 29	. 6 13 20 27 .	3 10 17 24 .	1 8 15 22 29
Thursday	2 9 16 23 30	. 7 14 21 28 .	4 11 18 25 .	2 9 16 23 30
Friday	3 10 17 24 .	1 8 15 22 29 .	5 12 19 26 .	3 10 17 24 31
Saturday	4 11 18 25 .	2 9 16 23 30 .	6 13 20 27 .	4 11 18 25 .

Exercise 3

Use the calendar above to help you to complete the sentences below.

> **Like this:**
> The last month of the year is ___December___ .

1. The first month of the year is _____ .
2. The sixth month of the year is _____ .
3. The tenth month of the year is _____ .
4. There are _____ months in a year.
5. October comes after the month of _____ .
6. December comes before the month of _____ .
7. April comes between _____ and _____ .
8. The middle two months of the year are _____ and _____
9. There are _____ days in the month of January.
10. The month of June has _____ days.
11. February is the only month which has exactly _____ days.
12. April, _____ , _____ and _____ all contain 30 days.
13. Bonfire Night is November 5th. On this calendar it is a _____ .
14. New Year's day is January 1st. On this calendar it is a _____ .
15. The first day of the week is a _____ .
16. The last day of the week is a _____ .
17. There are _____ days in a week.
18. Monday comes between _____ and _____ .
19. Saturday comes before _____ .
20. On this calendar my birthday is on a _____ .

Clocks

Exercise 4

Copy the table. Match up the times shown in the table with the times shown on the clock faces beneath. The first one is done for you.

Time	Clock
Eleven o'clock	A
Half past six	
Half past eight	
Six o'clock	
Quarter to six	

Time	Clock
Three o'clock	
Half past three	
Quarter to three	
Quarter to twelve	
Half past ten	

Time	Clock
Ten o'clock	
Quarter to ten	
Quarter past one	
Quarter past six	
Twelve o'clock	

Clock A   Clock B   Clock C   Clock D   Clock E

Clock F   Clock G   Clock H   Clock J   Clock K

Clock L   Clock M   Clock N   Clock O   Clock P

When telling the time, two types of clocks are used.
One type uses moving hands (analogue), the other uses changing figures (digital).

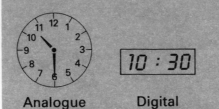

Analogue   Digital

The clocks look different but they both tell the same time.
(Half past ten.)

Exercise 5

Copy out the table below.
Match the times on the analogue and digital clocks with the times shown on the table.
The first one is done for you.

Time	Analogue	Digital
Quarter past one	1	*h*
Six o'clock		
Half past three		
Eight o'clock		
Nine o'clock		
Half past ten		
Two o'clock		
Quarter past six		
Quarter to one		
Quarter to three		

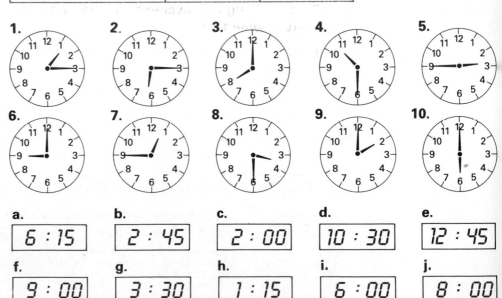

Barney's Day

Exercise 6

1. At what time does Barney ride to school?
2. When is Barney eating his breakfast?
3. At what time is Barney watching television?
4. At what time is Barney cleaning his teeth?
5. At what time is Barney eating school lunch?
6. At what time is Barney asleep?
7. At what time is Barney in his Maths lesson?
8. When do we see Barney with his friends?
9. What is Barney doing at 7.45 in the evening?
10. What is Barney doing at 8.30 in the morning?

Exercise 7

Choose the right answer to each question. If you think 'c' is the correct answer, write 1 : c.

1. How long is it between Barney brushing his teeth and Barney eating his breakfast?
   a. 2 hours.   b. 1 hour.   c. 15 minutes.

2. How long is it between Barney's Maths lesson and his eating lunch?
   a. 3 hours.   b. 2 hours.   c. 2 hours 15 minutes.

3. How much time passes between Barney eating his breakfast and Barney cycling to school?
   a. 1 hour.   b. 30 minutes.   c. 1 hour 30 minutes.

4. How much time passes between Barney's breakfast and his lunch?
   a. 3 hours 30 minutes.   b. 4 hours.   c. 5 hours 15 minutes.

5. How many hours pass between Barney's Maths lesson and his falling asleep?
   a. 12 hours.   b. 10 hours.   c. No time.

a.m. and p.m.

> One day lasts for 24 hours.
> During these 24 hours a time such as 10 o'clock will occur twice;
> 10 o'clock in the morning and 10 o'clock at night.
> If your friend says, 'I'll meet you at 7 o'clock', does your friend mean
> 7 o'clock in the morning or 7 o'clock in the evening?
> To help us, we use the terms a.m. and p.m.
>
> a.m.   The hours from 12 midnight, through the night and morning until 12 o'clock noon (12 hours).
>
> p.m.   The hours from 12 noon, through the afternoon and evening until 12 o'clock midnight.

Here are some things Barney does during one day.

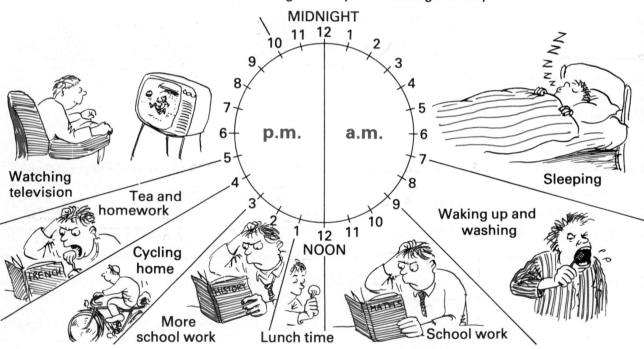

Exercise 8

Write out these sentences using a.m. or p.m. to complete the sentence.
1. Barney is cycling home at 3 (a.m. or p.m.).
2. Barney is watching television at 8 (a.m. or p.m.).
3. Barney is cleaning his teeth at 7 (a.m. or p.m.).
4. Barney is fast asleep at 3 (a.m. or p.m.).
5. Barney is doing his school work at 10 (a.m. or p.m.).
6. Barney finishes his lunch at 1 (a.m. or p.m.).
7. Barney starts homework and tea at 4 (a.m. or p.m.).
8. Barney goes to bed at 10 (a.m. or p.m.).
9. Barney starts his school day at 9 (a.m. or p.m.).
10. Barney starts his lunch break at 12 (noon or midnight).

Welcome to the Mad Hatter's Tea Party.

Everyone has a watch.
Each watch tells a different time.
Alice's watch tells the right time.

Remember this:
The Dormouse's watch is two hours slow.
The Mad Hatter's watch is one hour fast.
The March Hare's watch is fifteen minutes fast.

Exercise 9   Fill in a copy of the table below. The first two lines have been done for you.

Alice's Watch	Dormouse's Watch	Mad Hatter's Watch	March Hare's Watch
10.00	8.00	11.00	10.15
9.30	7.30	10.30	9.45
6.00			
8.00			
7.30			
11.30			
12.30			
5.45			
8.15			
6.20			
10.30			
7.40			
8.50			
2.05			
3.10			

Review 2

A. Mental arithmetic

1. How many 5p pieces are there in 25p?
2. What is 4 add 6, add 3, take away 7?
3. If you get 8 toffees in one box, how many will you get in 5 boxes?
4. How many days are there in 3 weeks?
5. What is 101 add 202?
6. How many 2p pieces are there in 26p?
7. If James saves 50p a day, how much would he save in 9 days?
8. What is 12 add 11, add 6, take away 5?
9. If Mary spends 72p, how much change will she get from £1.00?
10. Add one to ninety-nine.

B. Shape

1. Name these shapes.

a.   b.   c.   d.

2. Measure the radius of each circle.

a. 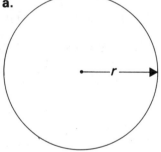   b. 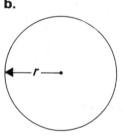   c.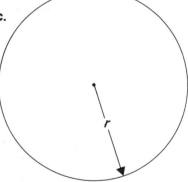

C. Symmetry

1. Match up the halves of these cars.

1   2   3   4   5   a   b   c   d   e

2. Complete the mirror image to make four words.

CODE BOOK HIDE BOX

D. Number work

1. Write out the answers.

a.	b.	c.	d.	e.
14 +23	26 +40	17 +29	207 +137	358 +248

f.	g.	h.	i.	j.
9 −6	34 −24	68 −40	32 −16	70 −44

k. $7 \times 2 =$ l. $5 \times 4 =$ m. $3 \times 6 =$ n. $8 \times 3 =$ o. $9 \times 2 =$

p. $2\overline{)8}$ q. $2\overline{)14}$ r. $2\overline{)24}$ s. $3\overline{)18}$ t. $3\overline{)39}$

2. Copy these and fill in the missing signs.

a. $4 * 2 = 6$ b. $3 * 3 = 9$ c. $7 * 4 - 11$ d. $10 * 4 = 6$

e. $5 * 5 = 25$ f. $8 * 2 = 4$ g. $8 * 1 = 7$ h. $14 * 4 = 18$

i. $10 * 2 = 5$ j. $4 * 4 = 16$ k. $6 * 9 = 15$ l. $12 * 12 = 0$

E. Number patterns

Fill in the missing numbers.

a. 2, 4, 6, ___, ___, 12   b. 4, 8, ___, 16, ___   c. 3, 6, ___, ___, 15

d. 3, 5, 7, ___, ___, ___   e. 7, 11, 15, ___, ___   f. 1, 6, 11, ___, ___

g. 9, ___, 17, 21, ___   h. 1, 7, 13, ___, ___   i. 2, 7, 12, ___, ___

j. 20, 18, 16, ___, ___   k. 18, 15, 12, ___, ___   l. 27, 25, ___, 21, ___

m. 31, 27, 23, ___, ___   n. 25, ___, 13, 7, ___   o. 23, 18, ___, 8, ___

F. Money

18p   29p   27p   24p   16p

1. How much change from 50p would you get if you bought a cup of tea?

2. How much change from 50p would you get if you bought a bar of chocolate?

3. How much change from 50p would you get if you bought a bottle of pop?

4. How much change from 50p would you get if you bought a comic?

5. How much change from 50p would you get if you bought a cake?

6. How much would a cake and a cup of tea cost together?

7. How much would a comic and a bar of chocolate cost?

8. How much would a bottle of pop and a comic cost?

9. How much would a comic and a cake cost?

10. How much would a cup of tea and a bar of chocolate cost?

G. Time

1. Match the written time with the correct clock time.

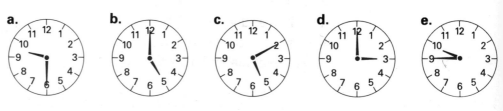

a.   b.   c.   d.   e.

Three o'clock   Half past nine   Five o'clock

Quarter to ten   Ten past five

2. If it is now 5 o'clock, what time will it be in five hours?

3. Does this clock show a quarter past or a quarter to five? $\boxed{4 : 45}$

4. If it is now half past two, what time will it be in five hours?

5. Does this clock show a quarter past or a quarter to six? $\boxed{6 : 15}$

6. If it is now 8 o'clock, what will the time be in $2\frac{1}{2}$ hours?

Exercise 1

Here are six orders. Write them out in Fred's code.
The first one is done for you.

1. A cup of tea and a sandwich = $t + s$.
2. A cup of tea and a slice of cake.
3. A glass of kola and a bun.
4. A bun and a sandwich.
5. A slice of cake and a bun.
6. A cup of tea and a glass of kola.

Exercise 2

Use Fred's menu to work out the amount of each bill.
The first one is done for you.

1. $t + c = 6 + 10 = 16$ pence   2. $t + k =$
3. $b + c =$ 4. $b + t =$
5. $f + c =$ 6. $s + f =$
7. $t + s =$ 8. $f + b + s =$
9. $c + s + t =$ 10. $s + s + t + k =$

Exercise 3

Here are six more orders. Write out the orders in Fred's code.
The first one is done for you.

1. 2 teas and a bun = $2t + b$ 2. 2 teas and one sandwich.
3. 1 kola and 2 cakes.   4. 1 tea, 1 sandwich and 2 buns.
5. 2 sandwiches and 3 teas.   6. 4 teas, 2 sandwiches and 3 cakes.

Exercise 4

Copy the table below and complete Fred's bills.

	order	bill
1.	$c + k = 10 + 17$	27 pence
2.	$c + b$	
3.	$r + s$	
4.	$c + t$	
5.	$t + r$	
6.	$2c = 2 \times 10$	20 pence
7.	$2s$	
8.	$3b$	
9.	$3r$	
10.	$t + 2b$	
11.	$f + 2c$	
12.	$2t + s$	
13.	$c + 2k$	
14.	$3c + 2t + f$	
15.	$2c + 3t + k$	
16.	$3s + 3t + 2b$	
17.	$2b + r + 2t$	
18.	$4r + t + 2k$	
19.	$r + 2t + k$	
20.	$2r + 2t + 2k$	

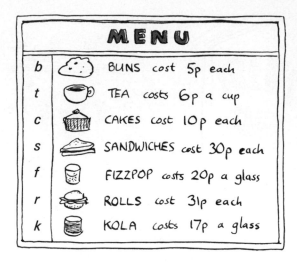

MENU

b BUNS cost 5p each
t TEA costs 6p a cup
c CAKES cost 10p each
s SANDWICHES cost 30p each
f FIZZPOP costs 20p a glass
r ROLLS cost 31p each
k KOLA costs 17p a glass

Exercise 5

Find the value of these expressions if:
$v = 2$ $w = 5$ $x = 1$ $y = 9$ and $z = 11$.

Like this: $x + y = 1 + 9 = 10$

1. $x + y$ **2.** $v + z$ **3.** $z + x$ **4.** $y + v$ **5.** $y + w$

6. $v + x + y$ **7.** $x + z + y$ **8.** $w + x + y$ **9.** $x + v + w$ **10.** $x + v + v$

11. $2v$ **12.** $2w$ **13.** $3y$ **14.** $2x + 2y$ **15.** $x + 3y$

16. $2x + 3y + v$ **17.** $2x + 4w$ **18.** $w + 2y$ **19.** $3v + x$ **20.** $2z + 3v$

The Great Magico knows that he has 6 rabbits altogether. He takes 2 rabbits from the hat. How many rabbits are left in the hat?

$$2 + * = 6$$
so, $* = 4$

Exercise 6

1. $4 + * = 6$	**2.** $2 + * = 6$	**3.** $5 + * = 9$	**4.** $3 + * = 7$
5. $5 + * = 11$	**6.** $9 + * = 11$	**7.** $7 + * = 9$	**8.** $5 + * = 15$
9. $9 + * = 14$	**10.** $10 + * = 20$	**11.** $20 + * = 25$	**12.** $* + 4 = 16$
13. $* + 3 = 6$	**14.** $* + 1 = 6$	**15.** $* + 11 = 13$	**16.** $x + 42 = 47$
17. $* + 1 - 11$	**18.** $* + 19 = 24$	**19.** $* + 14 = 23$	**20.** $25 + * = 50$
21. $23 + * = 63$	**22.** $* + 19 = 27$	**23.** $24 + * = 50$	**24.** $101 + * = 111$

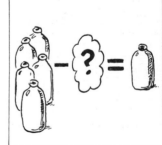

Magico puts five bottles in a box, and closes the lid.

When he opens the box there is only one bottle left.

How many bottles has Magico made disappear?

Exercise 7

Copy the questions. Find the number which should go into the box.

1. $6 - * = 4$	**2.** $7 - * = 4$	**3.** $9 - * = 7$	**4.** $8 - * = 5$
5. $10 - * = 4$	**6.** $13 - * = 7$	**7.** $18 - * = 9$	**8.** $20 - * = 15$
9. $30 - * = 21$	**10.** $35 - * = 29$	**11.** $50 - * = 30$	**12.** $50 - * = 32$

Exercise 8

Copy the questions. Find the number which is missing in each problem.

1. $* - 4 = 1$ 2. $* - 8 = 2$ 3. $* - 5 = 5$ 4. $* - 8 = 7$

5. $* - 6 = 10$ 6. $* - 7 = 7$ 7. $* - 3 = 11$ 8. $* - 10 = 11$

9. $* - 11 = 10$ 10. $* - 15 = 30$ 11. $* - 16 = 14$ 12. $* - 17 = 27$

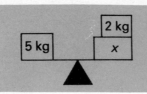

The scale is balanced.
So x must be 3 kg.

Exercise 9

1.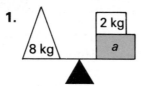

2.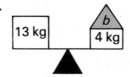

3.

6.

4.

5.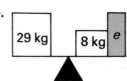

7.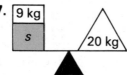

8.

9.

10.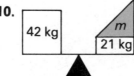

11.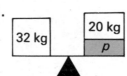

12.

13.

14.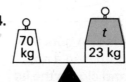

15.

16.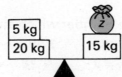

17.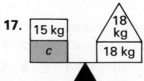

18.

Exercise 1

Use your ruler to measure these lines. Measure them to the nearest half centimetre.

1. ├────────────────────┤
2. ├──────────────────────────┤
3. ├──────────────────────────────┤
4. ├──────────┤
5. ├──────────────────────┤

Exercise 2

Use your ruler to draw these lines in your book.

1. 8 cm   **2.** 10 cm   **3.** 6 cm   **4.** 2 cm   **5.** 11 cm

Look at the drawings below.

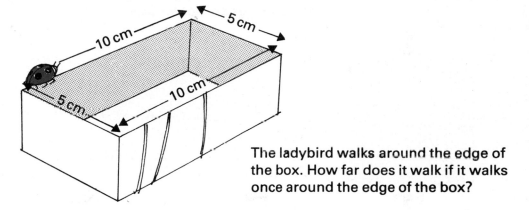

The ladybird walks around the edge of the box. How far does it walk if it walks once around the edge of the box?

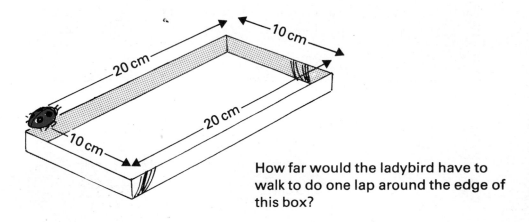

How far would the ladybird have to walk to do one lap around the edge of this box?

The distance around the edge of a shape is called the perimeter.

> To find the perimeter of a shape you add up the lengths of all the sides.

To find the perimeter of a rectangle, add the sides together.
Like this:

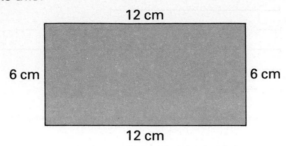

12 cm + 12 cm + 6 cm + 6 cm = 36 cm

The perimeter of this rectangle is 36 cm.

Exercise 3

Find the perimeter of these shapes.
These drawings are not full size.

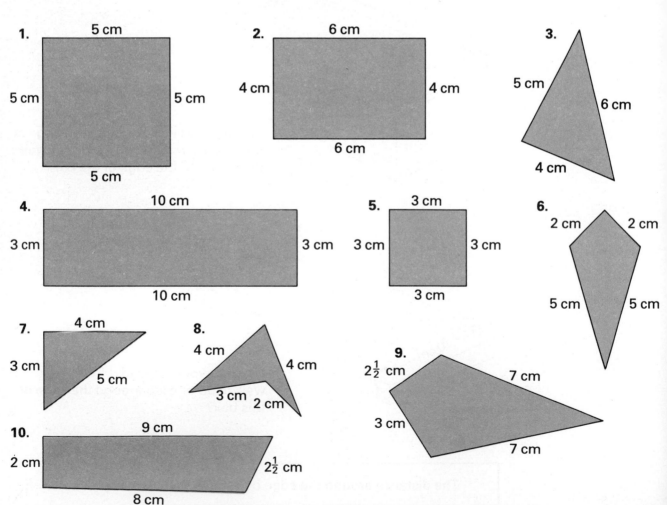

Look at the rectangle below. Its perimeter is 24 cm.
Try to find the length of the side marked with the letter *x*.

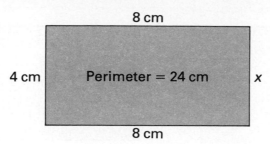

8 cm

4 cm   Perimeter = 24 cm   *x*

8 cm

Exercise 4

Find the length of the side marked with a letter in each drawing.
These drawings are not full size.

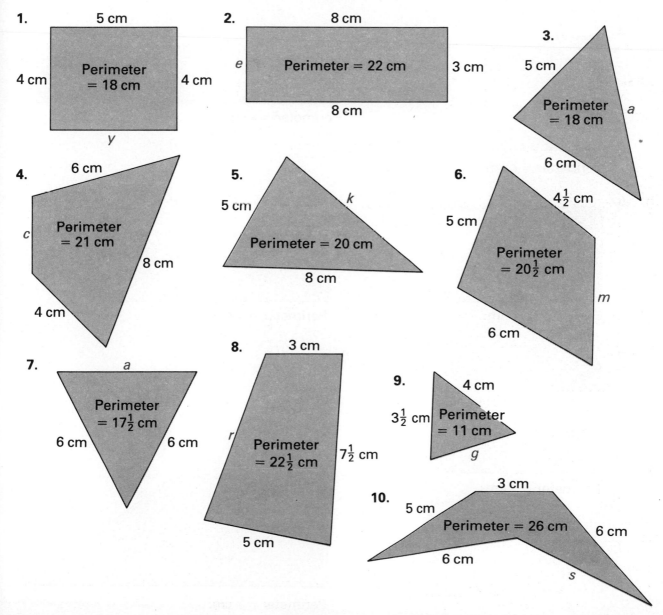

1.

5 cm

4 cm   Perimeter = 18 cm   4 cm

y

2.

8 cm

e   Perimeter = 22 cm   3 cm

8 cm

3.

5 cm

Perimeter = 18 cm   *a*

6 cm

4.

6 cm

c   Perimeter = 21 cm

8 cm

4 cm

5.

5 cm   *k*

Perimeter = 20 cm

8 cm

6.

$4\frac{1}{2}$ cm

5 cm

Perimeter = $20\frac{1}{2}$ cm

m

6 cm

7.

a

Perimeter = $17\frac{1}{2}$ cm

6 cm   6 cm

8.

3 cm

r   Perimeter = $22\frac{1}{2}$ cm $7\frac{1}{2}$ cm

5 cm

9.

4 cm

$3\frac{1}{2}$ cm   Perimeter = 11 cm

g

10.

3 cm

5 cm   Perimeter = 26 cm   6 cm

6 cm

s

page 81

Exercise 5   Measure these shapes. Draw them carefully. Work out the perimeter of each.

1.

Perimeter = ∗ cm.

2.

Perimeter = ∗ cm.

3.

Perimeter = ∗ cm.

4.

Perimeter = ∗ cm.

5.

Perimeter = ∗ cm.

6.

Perimeter = ∗ cm.

7.

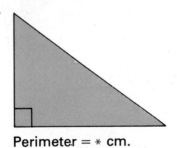

Perimeter = ∗ cm.

8.

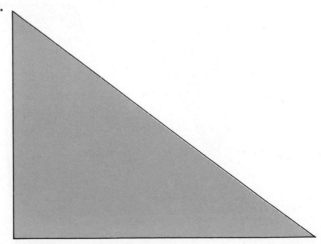

Perimeter = ∗ cm.

Watch what happens as the girl turns one arm of the fan.

The angle between the arms gets bigger.
The angle is the amount of turn between the two arms.

Here are six drawings showing the fan unfolding.
Under each drawing is a name. The name tells you what type of angle is shown by the fan.

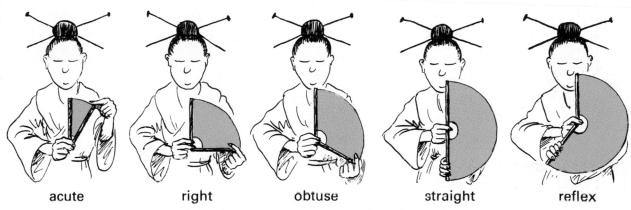

acute   right   obtuse   straight   reflex

Here are some more examples of angles.

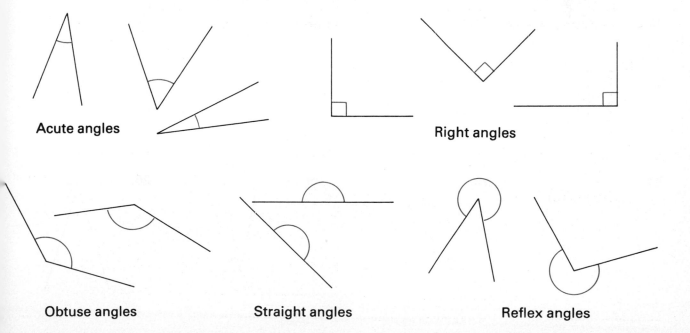

Acute angles   Right angles

Obtuse angles   Straight angles   Reflex angles

Exercise 1   Copy these angles into your book and name them.

1.
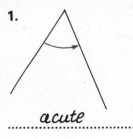

acute

2.
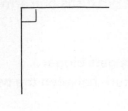

3.

4.

5.
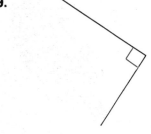

6.
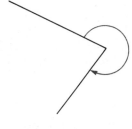

7.

8.
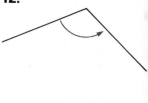

9.
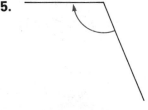

10.

11.

12.

13.
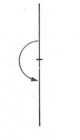

14.
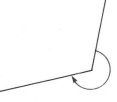

15.

16.
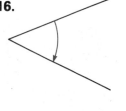

17.
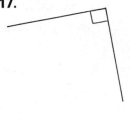

18.
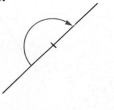

19.

20.

Choosing the right angle

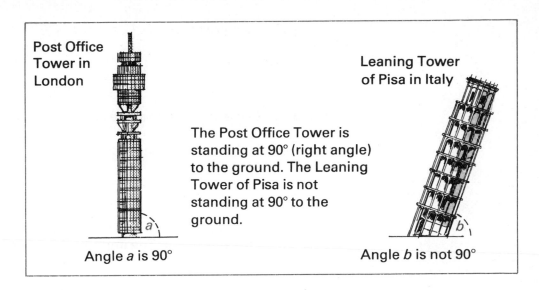

Post Office Tower in London

Leaning Tower of Pisa in Italy

The Post Office Tower is standing at 90° (right angle) to the ground. The Leaning Tower of Pisa is not standing at 90° to the ground.

Angle *a* is 90°

Angle *b* is not 90°

Exercise 2

Look at the drawings below. In each drawing an angle is marked with a letter. Copy the sentences and say whether they are true or false.

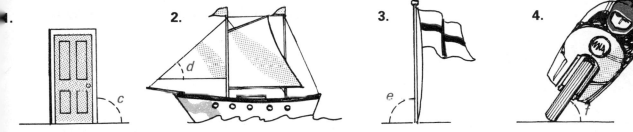

1.

Angle *c* is a right angle.

2.

Angle *d* is a right angle.

3.

Angle *e* is not a right angle.

4.

Angle *f* is a right angle.

5.

Angle *g* is a right angle.

6.

Angle *h* is not a right angle.

7.

Angle *j* is not a right angle.

8.

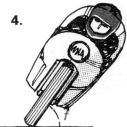

Angle *k* is a right angle.

9.

Angle *m* is not a right angle.

10.

Angle *n* is not a right angle.

To help us to draw the corners of squares and rectangles
we will now start to use a protractor.

The sides of a square must
come together at
the corners like this:

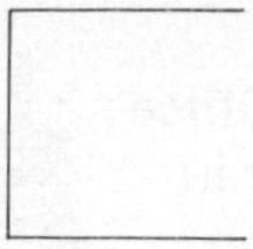

Not like this:

When the lines meet properly at the corners we call this a right angle.

How to draw a right angle using a protractor.

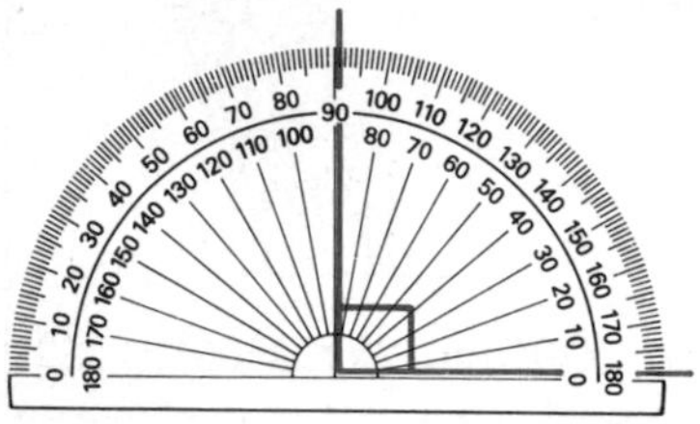

Here is a protractor. You can see a right angle marked
on the protractor. The important line is the 90° line.

Draw a base line. Make it about 6 cm long.

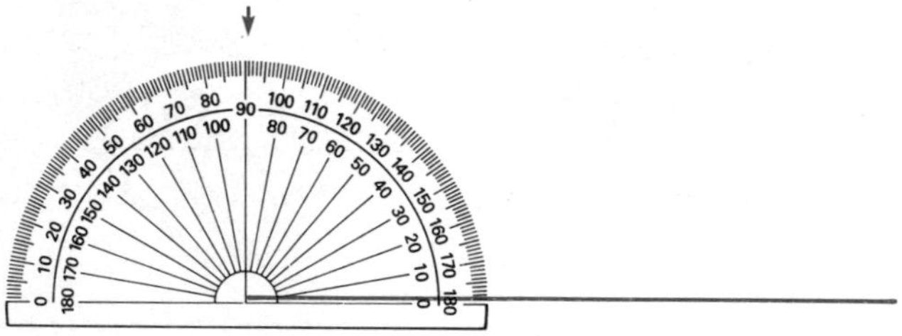

Put the centre of the protractor on one end of the
base line.

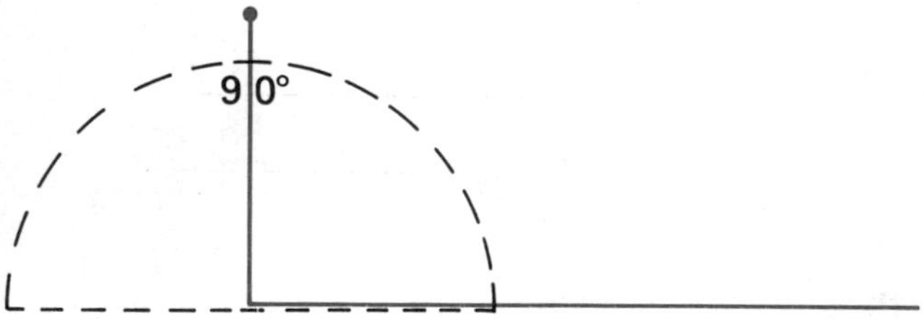

Find the 90° line on the protractor, mark it and
join this point to the end of the base line.
You have now drawn a right angle.

Exercise 3   Draw some right angles, like the ones below.

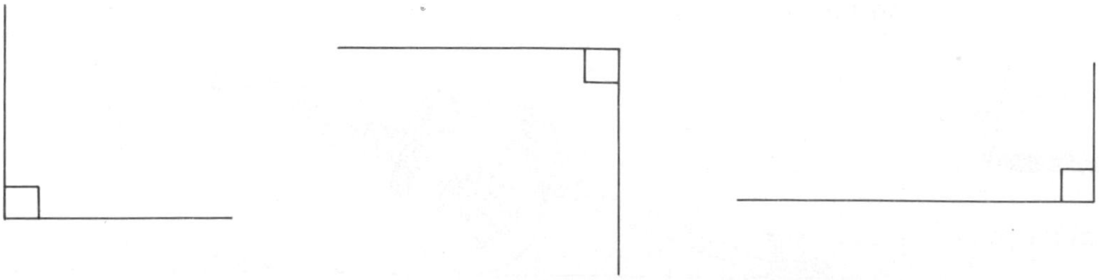

Exercise 4

Draw these lines in your book.
The drawings are not shown full size.

1.

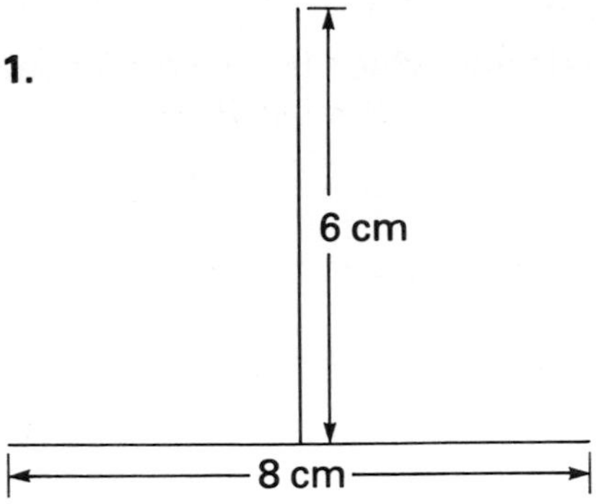
6 cm
8 cm

2.

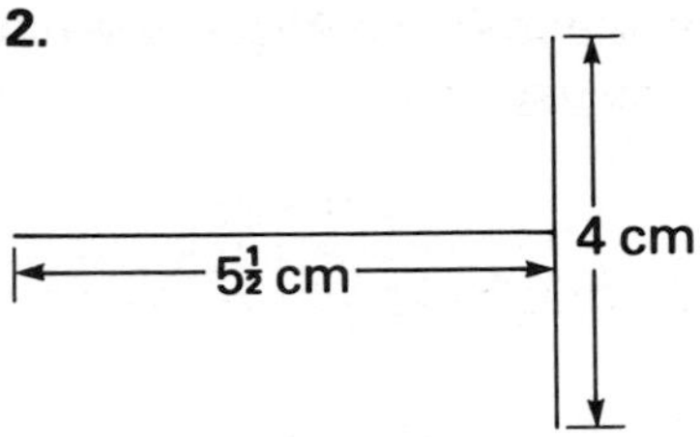
4 cm
5½ cm

3.

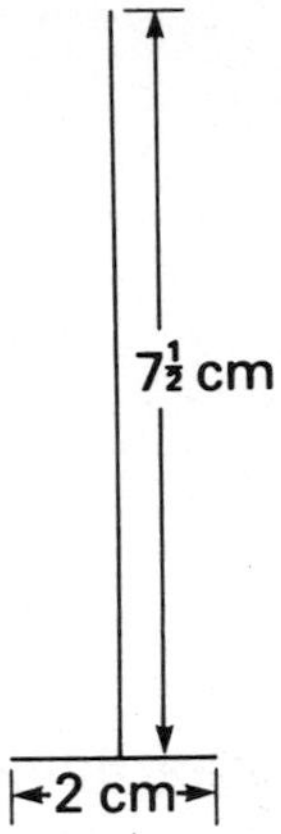
7½ cm
2 cm

Exercise 5

Look at these angles. Decide if each angle is 90° (right angle), more than 90° or less than 90°.

1.

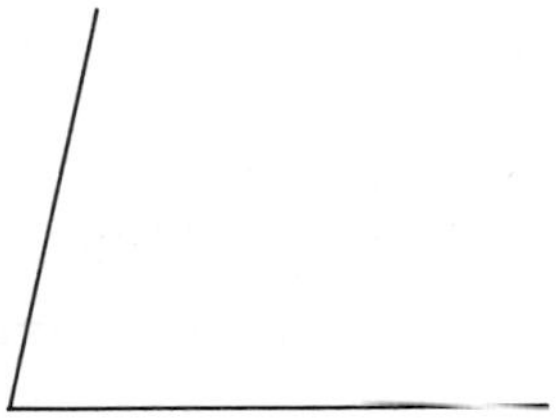

2.

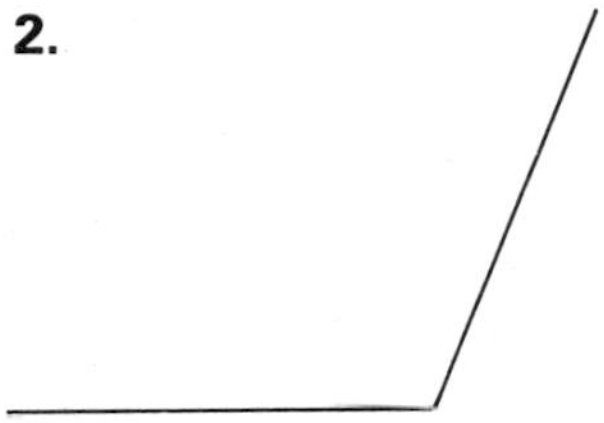

3.

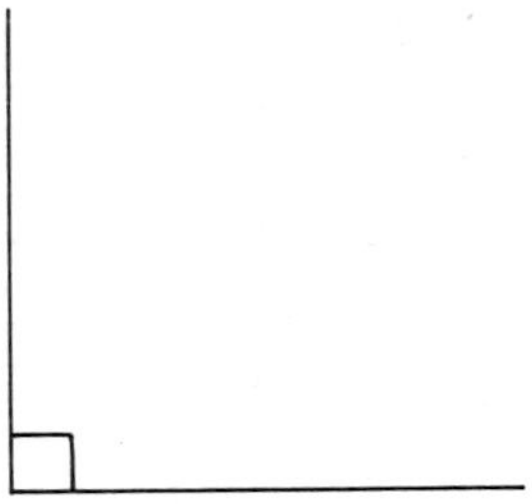

4.

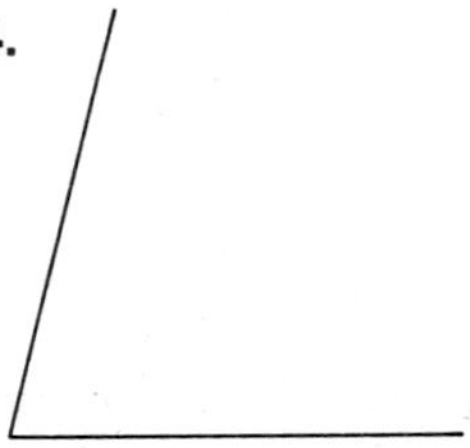

5.

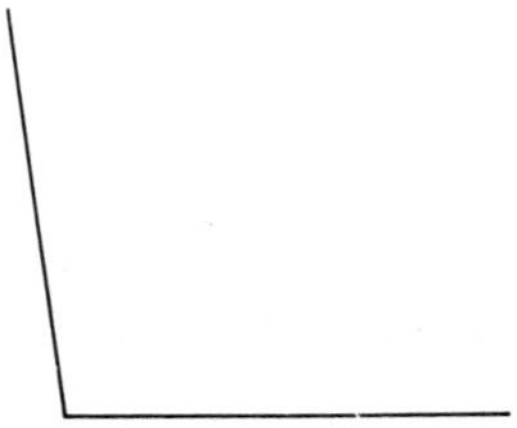

6.

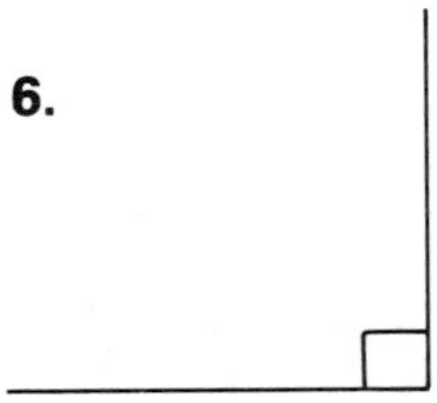

7.

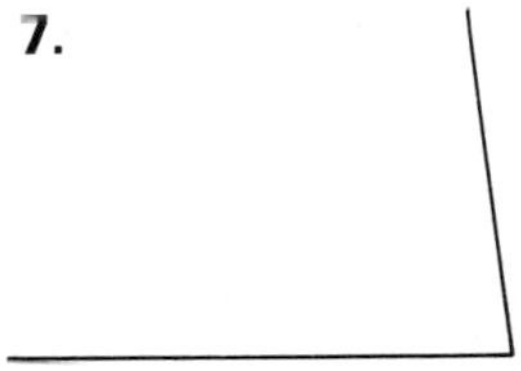

8.

9.

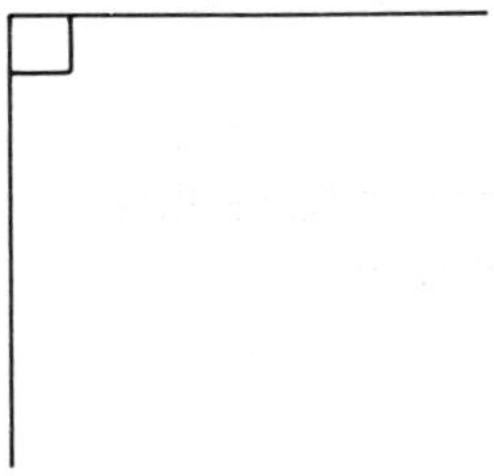

10.

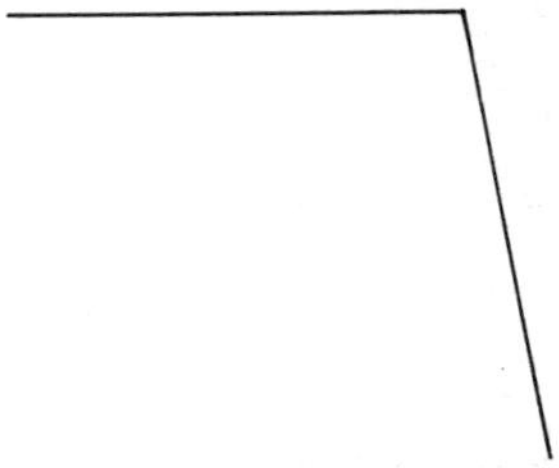

Copy out this table and put the number of each angle in the correct column.

more than 90°	right angle 90°	less than 90°

11.

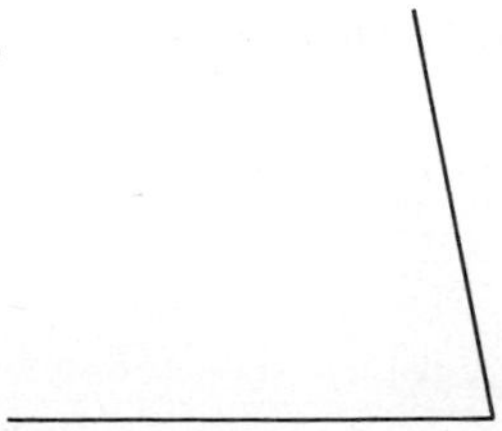

12.

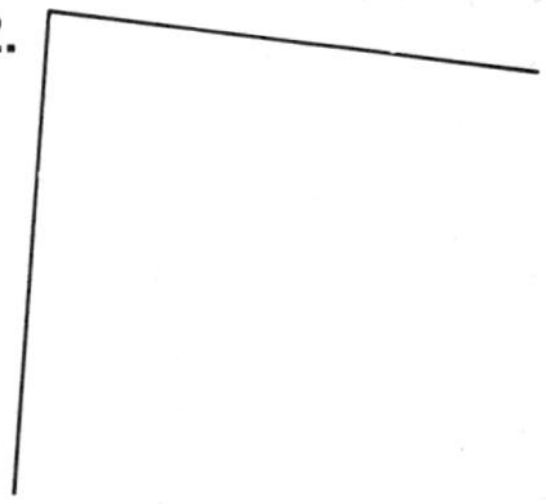

If you look at your protractor, you will see that there are two sets of numb
One set goes from 0° up to 180°.
The other set goes from 180° down to 0°.

When we measure an angle, we have to decide which is the correct set of numbers to use.

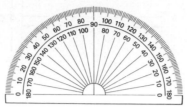

Measuring angles

Here are two angles. You will see that the angles are at different ends of the baseline.

When we place the protractor on the angle, we will use the set of numbers that start at 0°.

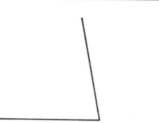

Start at 0°
Not at 180°

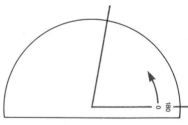

Start at 0°
Not at 180°

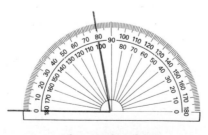

An angle of 80°

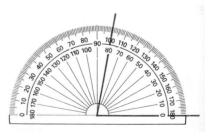

An angle of 80°

Angles are measured in degrees. This small
circle ° above a number means 'so many degrees'.
Example, 45 degrees is written 45°.

xercise 6   Measure these angles.

1.
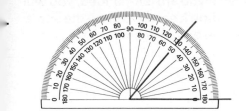

2.
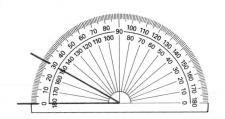

3.
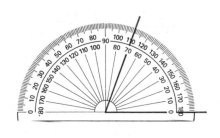

4.
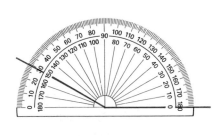

5.
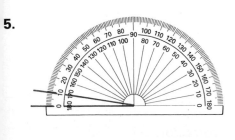

6.
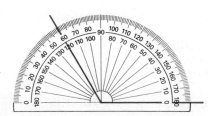

7.
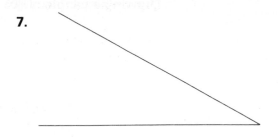

8.
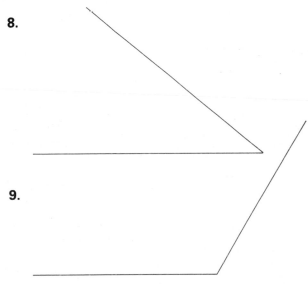

9.
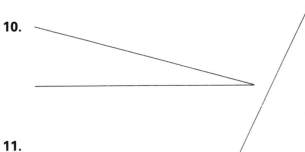

10.

11.
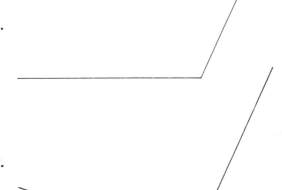

12.

Drawing angles

Drawing an angle of 60°.

If you are left-handed, you will find it easier to draw the angle in this direction.

a.

Draw a line about 6 cm long.

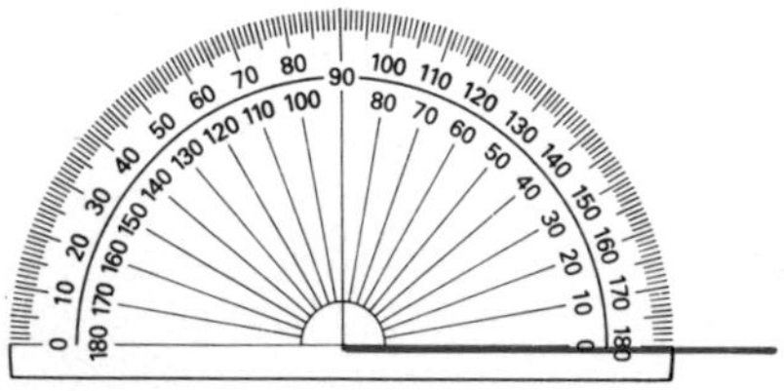

b.

Place the centre of the protractor at one end of the line.

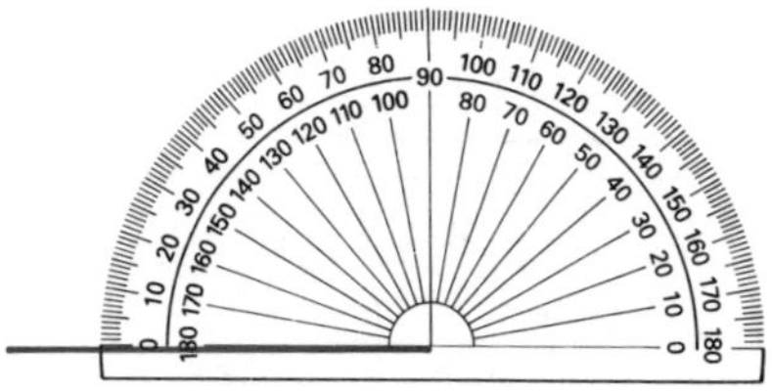

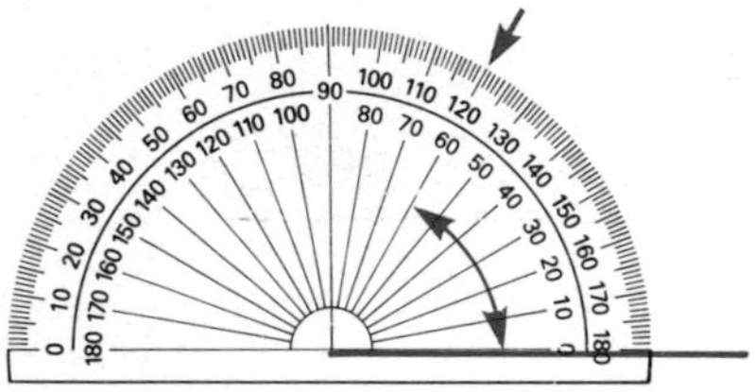

c.

Find the 60° mark on the correct scale. Carefully mark the paper.

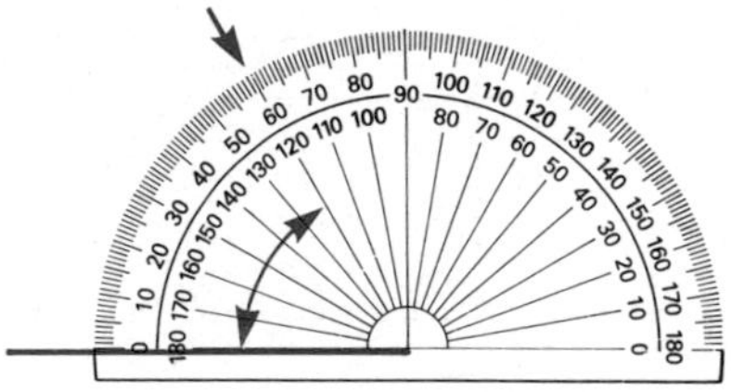

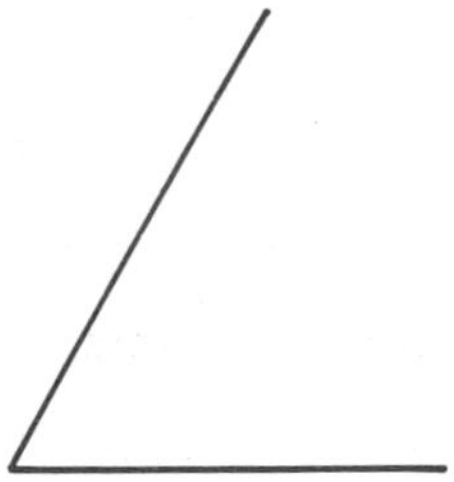

d.

Now join this mark to the end of the line.

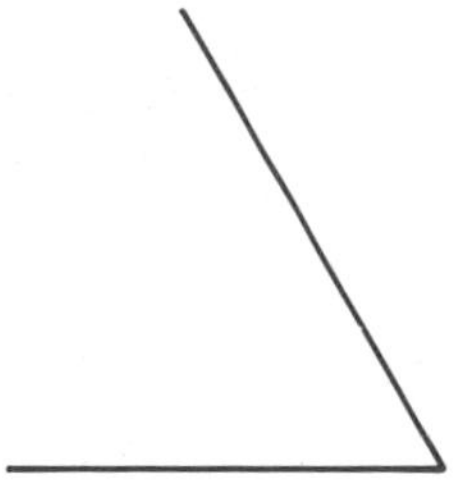

Exercise 7

Draw these angles in your book, using a protractor.

1. 30°	**2.** 50°	**3.** 70°	**4.** 90°
5. 100°	**6.** 130°	**7.** 45°	**8.** 35°
9. 65°	**10.** 125°	**11.** 145°	**12.** 105°
13. 15°	**14.** 115°	**15.** 10°	**16.** 135°

Review 3

Measurement   Measure the height of these chess pieces.

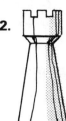

1.

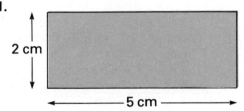

2.

3.

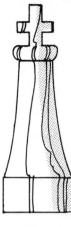

4.

Pawn   Rook   Bishop   King

B. Drawing   Draw these shapes

1.

2 cm

5 cm

2.

3 cm

2 cm

2 cm

4 cm

C. Sorting   Sort these words into four groups:
colours, first names, shapes or numbers.

Tom	blue	five	yellow	circle	rectangle
square	six	Mary	John	nine	green
red	four	brown	seven	Ann	triangle

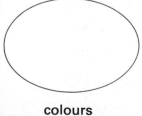

colours

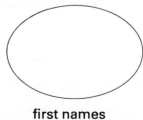

first names

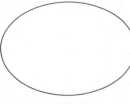

shapes

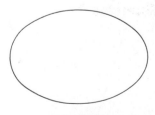

numbers

D. Greater than/less than

> means greater than   < means less than

Copy out these problems and put in the correct sign.

1. 2 * 4
2. 5 * 6
3. 9 * 5
4. 9 * 8
5. 2 + 3 * 8
6. 9 + 4 * 12
7. 6 + 7 * 14
8. 8 + 9 * 19
9. 8 − 7 * 3
10. 12 − 5 * 9
11. 14 − 7 * 8
12. 21 − 10 * 9

E. Algebra

Copy these out and fill in the missing numbers.

1. 6 + * = 8
2. 8 + * = 10
3. 6 + * = 10
4. * + 7 = 11
5. * + 9 = 13
6. * + 8 = 17
7. 9 − * = 4
8. 12 − * = 9
9. 18 − * = 12
10. * − 5 = 3
11. * − 2 = 5
12. * − 8 = 9

F. Perimeter

Calculate the length of these perimeters.
Remember: the perimeter of a shape is the distance once around the edge

1.

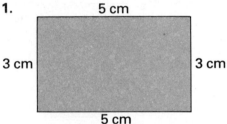

5 cm
3 cm   3 cm
5 cm

2.

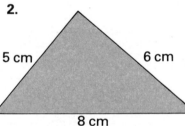

5 cm   6 cm
8 cm

3.

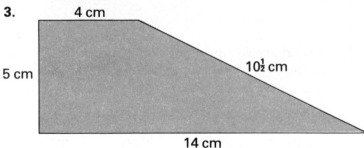

4 cm
5 cm   10½ cm
14 cm

G. Angles

Name these angles.

1.

2.

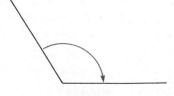

3.

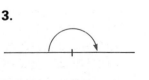

4.

Time

a.m.
The hours from
12 o'clock midnight
to 12 o'clock midday.

p.m.
The hours from
12 o'clock midday
to 12 o'clock
midnight.

1.  3 p.m.

    a. What time will it be in 2 hours?
    b. What time will it be in 6 hours?
    c. What time will it be in 8 hours?

2. 8 a.m.

    a. What time will it be in $2\frac{1}{2}$ hours?
    b. What time will it be in 3 hours?
    c. What time will it be in $3\frac{1}{2}$ hours?

3. 1.30 p.m.

    a. What time will it be in 6 hours?
    b. What time will it be in $2\frac{1}{2}$ hours?
    c. What time will it be in $4\frac{1}{2}$ hours?

4. _____ comes between Thursday and Saturday.
5. _____ comes between Sunday and Tuesday.
6. _____ comes after Saturday.
7. _____ comes after Tuesday.
8. _____ comes before Friday.

Money

Tennis racket
£6.20

Soccer ball
£8.50

Track suit
£15.25

Rugby boots
£12.00

1. If you bought the rugby boots, how much change would you get from £20?
2. If you bought the soccer ball, how much change would you get from £10?
3. How much would the tennis racket and the rugby boots cost together?
4. How much would the track suit and the soccer ball cost together?
5. How much would the track suit and the tennis racket cost together?
6. If the cost of the track suit was reduced by £4, how much would it cost?
7. If the rugby boots were reduced by £2.50, how much would they cost?
8. How much would two track suits cost?

Dividing numbers into ten

The Decland Boys have been given a big bar of chocolate. They have a problem. They must share out the chocolate equally. Each boy must have the same size of piece.

Count the number of boys. How many pieces must be broken?

When the boys open the wrapper, they see that the bar has already been broken into ten pieces.

Each piece is 'one part of ten' because ten pieces make up a whole-one.

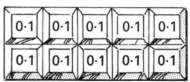

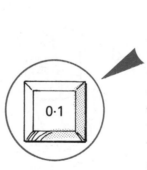

On each of the ten pieces the number 0·1 is written. This is the decimal way of writing 'one part of ten'.

Two pieces would be two parts of ten, written 0·2.

Exercise 1

Write out these numbers as decimals.

1. One part of ten
2. Two parts of ten
3. Six parts of ten
4. Eight parts of ten
5. Four parts of ten
6. Seven parts of ten
7. Three parts of ten
8. Five parts of ten
9. Nine parts of ten

Exercise 2

Copy and complete these sentences.

1. 0·9 is the same as _____ tenths
2. 0·5 is the same as _____ tenths
3. 0·1 is the same as _____ tenth
4. _____ tenths is the same as 0·3
5. 0·4 is the same as _____ tenths
6. _____ tenths is the same as 0·7

Copy the two headings below and answer the two questions for each drawing.

	What decimal part is missing?	What decimal part is left?
	0·1	0·9

10.

11.

12.

13.

14.

15.

16.

17.

18.

7.

8.

9.

Exercise 4

For each rod   **a.** Say what part of the rod is coloured.
                   b. Say what part of the rod is uncoloured.

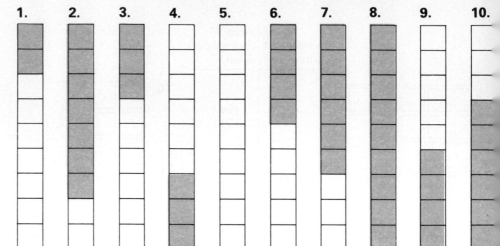

1.   2.   3.   4.   5.   6.   7.   8.   9.   10.

11.   12.   13.   14.   15.   16.   17.   18.   19.   20.

Exercise 5

Say what part of each circle is   **a.** coloured   **b.** uncoloured.

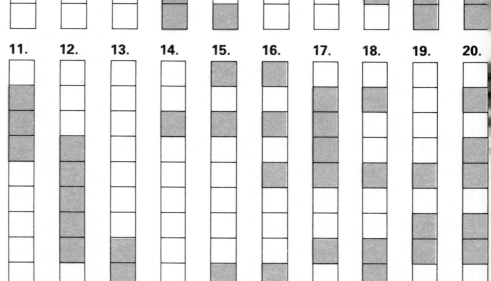

1.   2.   3.   4.   5.

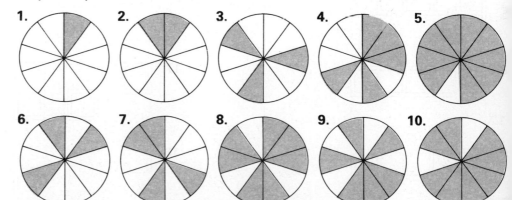

6.   7.   8.   9.   10.

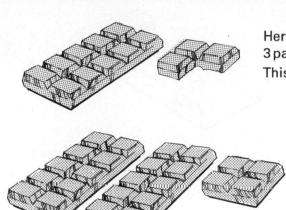

Here is 1 whole bar and
3 parts out of ten.
This is written 1·3

Here are 2 whole bars and
4 parts out of ten.
This is written 2·4

The decimal point separates the whole ones (units)
from the part out of ten (tenths).

Exercise 6   Write down the number of bars of chocolate shown in each picture.
Remember to make the decimal point clear.

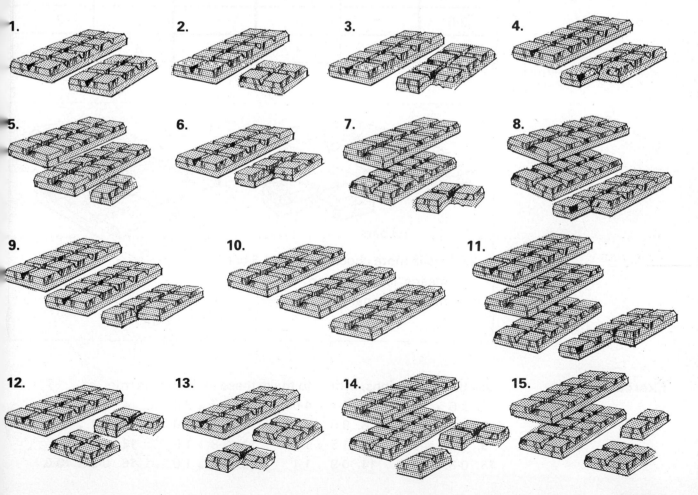

1.

2.

3.

4.

5.

6.

7.

8.

9.

10.

11.

12.

13.

14.

15.

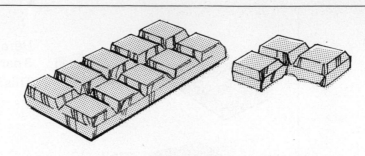

Here are 1·3 bars of chocolate.
1·3 means 1 'whole' and 3 'tenths'.

Exercise 7

Copy the table below. Fill in the missing spaces.

Number	Units	Tenths	Number	Units	Tenths
1·3	1	3	5·0	5	*
1·9	1	*	7·5	*	5
2·7	*	7	*	8	7
3·6	*	6	*	9	2
9·2	9	*	*	6	9

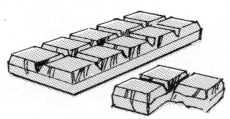

1·3 bars

2·0 bars

2·0 bars is more chocolate than 1·3 bars.
> means 'is bigger than'
< means 'is smaller than'

So, 2·0 > 1·3
or 1·3 < 2·0

Exercise 8

Use these two signs, > or < to make these expressions true.

1. 3·6 * 1·6   2. 5·1 * 4·0   3. 4·7 * 3·8   4. 9·0 * 6·5
5. 5·7 * 6·0   6. 8·5 * 9·3   7. 1·7 * 1·6   8. 8·2 * 8·3
9. 4·7 * 4·9   10. 1·5 * 1·0   11. 3·8 * 1·9   12. 9·6 * 9·0
13. 0·3 * 0·1   14. 0·9 * 1·1   15. 0·7 * 1·0   16. 0·9 * 10·0

Decimals – addition and subtraction

If you add,  you get ⬚ which is 0·3

You can add decimal parts, just like whole numbers.
Remember the decimal point.   So 0·1 + 0·1 + 0·1 = 0·3

Exercise 9

Add up these decimal numbers. Remember the decimal point.

	1.		**2.**		**3.**		**4.**		**5.**
	0·1		0·2		0·6		0·4		0·7
	+0·7		+0·2		+0·3		+0·1		+0·2

	6.		**7.**		**8.**		**9.**		**10.**
	0·3		0·8		0·2		0·7		0·5
	+0·3		+0·1		+0·4		+0·1		+0·2

Exercise 10

These two amounts (0·3 and 0·7) add up to 1·0 which is 1 whole.
Copy the questions into your book. Add up the numbers.

	1.		**2.**		**3.**		**4.**		**5.**
	0·3		0·2		0·4		0·5		0·3
	+0·7		+0·8		+0·2		+0·5		+0·6

	6.		**7.**		**8.**		**9.**		**10.**
	0·5		0·6		0·1		0·1		0·2
	+0·3		+0·4		+0·9		+0·7		+0·7

Look at your answers. Put a star by the answers which add up to exactly
1 whole.

You can add decimals and whole numbers as
easily as the addition above.

Like this:
```
  2.3
 +3·1
 ────
  5·4
```

Exercise 11

	1.		**2.**		**3.**		**4.**		**5.**
	2·3		3·1		6·6		4·5		0·3
	+1·4		+2·5		+2·3		+2·3		+2·1

	6.		**7.**		**8.**		**9.**		**10.**
	6·4		7·2		1·9		4·8		9·7
	+3·3		+1·6		+6·0		+3·1		+0·2

0·3 bars

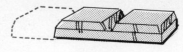

0·2 bars

Subtraction of decimals is as easy as addition.
In the pictures above, 0·1 has been taken from 0·3.
0·2 is left.
So, 0·3 − 0·1 = 0·2

Exercise 12

Subtract these decimal numbers. Remember to show the decimal point.

1. 0·6 −0·2	**2.** 0·7 −0·4	**3.** 0·5 −0·1	**4.** 0·9 −0·3	**5.** 0·7 −0·6
6. 0·8 −0·3	**7.** 0·4 −0·2	**8.** 0·9 −0·7	**9.** 0·8 −0·1	**10.** 0·9 −0·9

Exercise 13

Subtract these decimal numbers. Remember to show the decimal point.

1. 1·5 −1·2	**2.** 1·7 −1·2	**3.** 1·6 −0·5	**4.** 1·9 −0·7	**5.** 2·8 −1·3
6. 6·2 −2·1	**7.** 5·7 −3·3	**8.** 4·1 −2·1	**9.** 9·6 −7·4	**10.** 7·9 −4·5

Exercise 14

Subtract these decimal numbers. Remember to show the decimal point.

1. 12.6 −1·5	**2.** 23·6 −2·4	**3.** 13·9 −1·8	**4.** 35·6 −21·3	**5.** 45·7 −21.5
6. 39·9 −27·8	**7.** 55·5 −21·4	**8.** 62·5 −31·5	**9.** 49·4 −26·4	**10.** 95·1 −72·0

Exercise 15

Subtract these decimal numbers. These are harder questions.
Remember to show the decimal point.

1. 9·6 −2·7	**2.** 3·2 −1·9	**3.** 4·1 −2·5	**4.** 6·5 −4·7	**5.** 7·2 −5·4
6. 6·3 −4·8	**7.** 8·0 −4·5	**8.** 6·0 −3·7	**9.** 5·0 −4·9	**10.** 7·3 −6·9

More work with circles

The goat is eating the grass. When the chain becomes tight the goat cannot go any further, so it moves around a little. When the goat has eaten all that it can, what shape will it have made?

Exercise 1

1. Trace the path of the goat like this.

   a. Put a mark near the middle of page and mark it C for centre.

   b. Mark another point 3 cm from C.

   c. Keep marking many more points all 3 cm away from point C.

   d. If you draw enough points you will draw the same shape of path as the goat made.

   e. Join up the points with a curve and say what shape you have drawn.

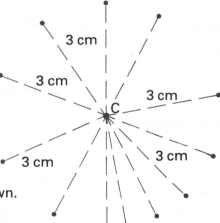

2. Now try to do this again using a distance of 5 cm this time.

The shapes which you should have drawn are both circles.

A circle is a path which always stays the same distance from a centre point.

Parts of a circle

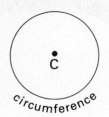
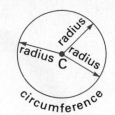
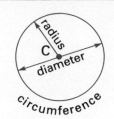

The circumference is the distance around the edge of the circle.

The radius is the distance from the centre to the edge of the circle.

The diameter is the distance all the way across the circle. The diameter must go through the centre.

Exercise 2

Measure **a.** the radius, **b.** the diameter of each of the five circles below.

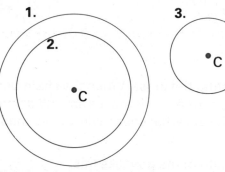

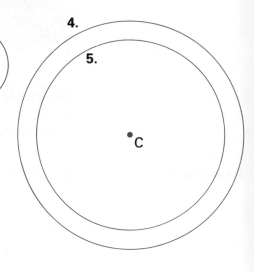

Copy and complete these sentences.
6. The diameter is _____ the length of the radius.
7. The radius is _____ the length of the diameter.

Exercise 3

Copy and complete the sentences below. The circles are *not* drawn to scale.

1. The diameter of this circle is _____ cm.

2. The radius of this circle is _____ cm.

3. The radius of this circle is _____ cm.

4. The diameter of this circle is _____ cm.

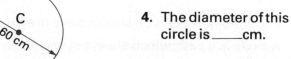

5. The radius of this circle is ____ cm.

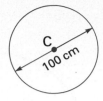

6. The diameter of this circle is ____ cm.

7. The diameter of this circle is ____ cm.

8. The diameter of this circle is ____ cm.

Exercise 4

Find the missing measurements. Copy and complete the sentences. The drawings are not drawn to scale.

1. Each circle is the same size.
The diameter is ____ cm.
The radius is ____ cm.

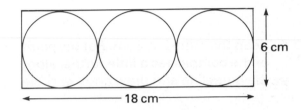

2. Each circle is the same size.
The diameter is ____ cm.
The radius is ____ cm.

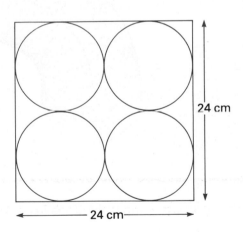

3. The height *h* of the cross is ____ cm.
The width *w* of the cross is ____ cm.

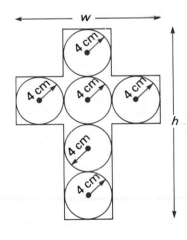

4. The diameter of the large circle is ____ cm.
The radius of the large circle is ____ cm.
The diameter of each small circle is ____ cm.
The radius of the small circles is ____ cm.

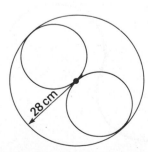

Patterns with circles

Exercise 5

1. Draw a curved guideline with a pair of compasses using a radius of about 3 cm, like this:

2. Now reset your compasses to a radius of 2 cm.

3. Put the point of the compasses on one end of the guideline and draw a full circle.

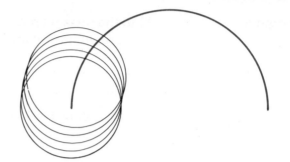

4. Keep the radius at 2 cm. Put the point of the compasses a little further along the guideline and draw another circle.

5. Keep drawing the 2 cm circles until this 'spring' pattern is complete.

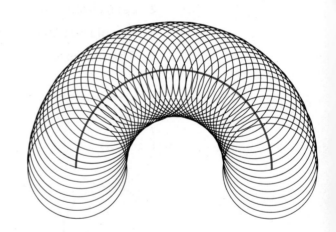

Exercise 6

Repeat this type of pattern using guidelines like those shown below.

1.

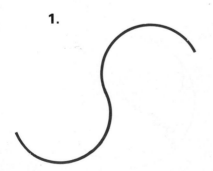

2.

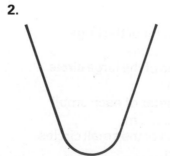

3.

Exercise 7

Here is a way to make another type of
pattern with circles.

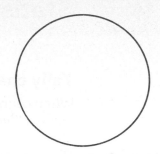

1. In the middle of a page in your book, draw
   a circle of radius $2\frac{1}{2}$ cm.

2. Keep the compass gap at $2\frac{1}{2}$ cm. Choose
   a point P on the circumference.
   Put the compass point on P and draw
   a circle.

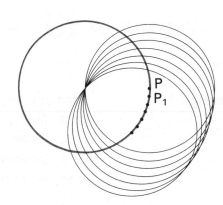

3. Move the point of the compasses around
   the circle a little to P_1.
   Draw another circle having the same radius.

4. Move the point of the compass again and
   draw another circle.
   Draw more circles until the pattern is
   complete.

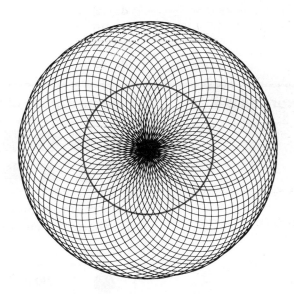

Statistics

Tally charts

When counting a group of items, it is sometimes useful to keep a 'tally' using a 'five-bar gate'. For each item in the group, you make one mark.

Here are 3 dots As you count each one you make a mark. III = 3

Here are 5 dots This is shown as ℍℍ = 5

Here are 9 dots This is shown as ℍℍ IIII = 9

Notice that each group of 5 is shown like this: ℍℍ

Exercise 1

What numbers are shown here?

1. IIII = 4
2. ℍℍ =
3. ℍℍ II =
4. ℍℍ ℍℍ =
5. ℍℍ ℍℍ III =
6. ℍℍ ℍℍ ℍℍ =
7. ℍℍ ℍℍ ℍℍ ℍℍ =
8. ℍℍ ℍℍ ℍℍ IIII =
9. ℍℍ ℍℍ I =
10. ℍℍ ℍℍ IIII =
11. ℍℍ ℍℍ ℍℍ III =
12. ℍℍ I =
13. ℍℍ III =
14. ℍℍ ℍℍ ℍℍ ℍℍ IIII =
15. ℍℍ ℍℍ ℍℍ II =

Exercise 2

Write these numbers down on 'five-bar gates'.

1. **14** = ℍℍ ℍℍ IIII
2. 7
3. 12
4. 16
5. 14
6. 17
7. 20
8. 22
9. 15
10. 23
11. 19
12. 13
13. 25
14. 27
15. 31

Exercise 3

Here are a group of shapes.

Copy the tally chart below. As you count each type of shape, make a mark on the chart. The circles have been done for you.

shape	tally	total
●	ℍℍ II	7
■		
▲		
▬		

Exercise 4

From the tally chart in exercise 3 we can make a display.
Here is the completed tally chart, and a pictogram displaying the information.

shape	tally	total
●	ⅢⅢ II	7
■	ⅢⅢ ⅢⅢ	10
▲	ⅢⅢ II	7
▬	ⅢⅢ	5

pictogram

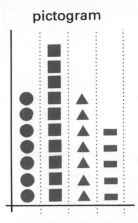

1. How many circles are there?
2. How many squares are there?
3. How many triangles are there?
4. How many rectangles are there?
5. How many shapes are there in all?

Exercise 5

Tally this new group of shapes. Copy the tally chart and pictogram and complete both.

shape	tally	total
◆	III	3
●		
▲		
■		

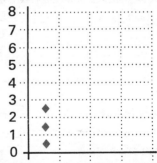

Exercise 6

Here is a group of vehicles found parked in a street.
Copy the tally chart and pictogram and complete both.

vehicle		tally	total
Cars	🚐	ⅢⅢ I	6
Motorbikes	🛵		
Lorries	🚚		
Bicycles	🚲		

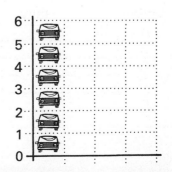

Exercise 7

People walking past the school gate were asked to vote for their favourite colour. Here are the results.

Red	Blue	Red	Yellow	Blue	Green	Red	Green	Brown
Green	Red	Blue	Pink	Yellow	Blue	Blue	Blue	Green	Red
Blue	Brown	Yellow	Green	Blue	Red	Blue	Brown	Blue

1. Copy and complete the tally chart below.

2. Copy and complete the pictogram.

colour	tally	total
Red	ⅡⅡ Ⅰ	6
Blue		
Yellow		
Green		
Brown		
Pink		

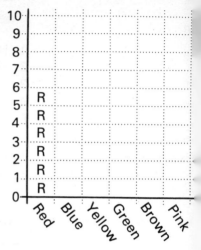

3. Which was the most popular colour?

4. Which colour got least votes?

5. Which colours got the same number of votes?

6. Which colour got 5 votes?

7. How many people voted?

Exercise 8

Here are the maths grades given to class 1B.

B –	C –	B –	C –	D –	E –	C –	C –	C
A –	D –	F –	B –	C –	D –	B –	C –	A
E –	B –	C –	E –	C –	D –	F –	A –	C

1. Copy and complete the tally chart below.

2. Copy and complete the pictogram.

number	tally	total
A	Ⅲ	3
B		
C		
D		
E		
F		

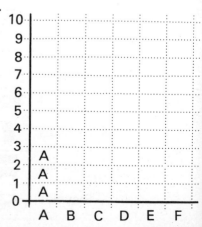

Exercise 9

Here are five items that are sold in the school tuck shop.

Cola   Crisps   Toffees   Chocolate   Chews

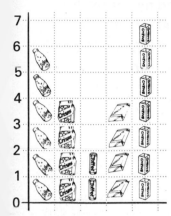

Answer these questions about the pictogram on the left. It shows how many of each item were sold during break time.

1. How many packs of toffees were sold?
2. How many bottles of cola were sold?
3. How many bags of crisps were sold?
4. How many bars of chocolate were sold?
5. Which item sold most?
6. Which item sold least?
7. Which two items sold the same amount?
8. How many items were sold in all?

Exercise 10

When a class voted for their favourite sweets, this was the result.

Lolly pop   Ice Cream   Chocolate   Toffees   Bubble Gum

4 voted for Lollypops – 𝑘 𝑘 𝑘 𝑘

7 voted for Ice Cream – 𝑘 𝑘 𝑘 𝑘 𝑘 𝑘 𝑘

6 voted for Chocolate – 𝑘 𝑘 𝑘 𝑘 𝑘 𝑘

5 voted for Toffees – 𝑘 𝑘 𝑘 𝑘 𝑘

3 voted for Bubble Gum 𝑘 𝑘 𝑘

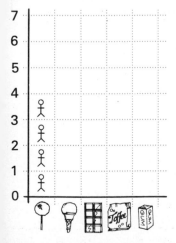

Copy and complete the pictogram shown on the left.
Then answer these questions.

1. How many pupils voted for bubble gum?
2. How many pupils voted for ice cream?
3. How many pupils voted for toffees?
4. How many pupils voted for chocolate?
5. Which item got most votes?
6. Which item got least votes?
7. Which item got 6 votes?
8. Which item got 4 votes?
9. Which item got 5 votes?
10. How many votes were cast in all?

Exercise 11

If we asked the whole school to vote for their favourite sweet, our pictogram would be very big and hard to understand.

So for this pictogram we will use one figure (♀) to show each 10 votes.

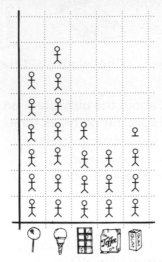

♀ = 10 pupils

♀ = 5 pupils

1. How many pupils voted for lollypops?
2. How many pupils voted for ice cream?
3. How many pupils voted for chocolate?
4. How many pupils voted for toffees?
5. How many pupils voted for bubble gum?

Remember each figure (♀) = 10 votes.

Exercise 12

Here is a tally chart of how pupils make their way to school.

type of transport	tally	total
Bus	ⅢⅢ ⅢⅢ ⅢⅢ ⅢⅢ ⅢⅢ ⅢⅢ ⅢⅢ ⅢⅢ	
Walk	ⅢⅢ ⅢⅢ ⅢⅢ ⅢⅢ ⅢⅢ ⅢⅢ ⅢⅢ ⅢⅢ ⅢⅢ ⅢⅢ ⅢⅢ	
Cycle	ⅢⅢ ⅢⅢ ⅢⅢ ⅢⅢ ⅢⅢ ⅢⅢ	
Train	ⅢⅢ ⅢⅢ	
Car	ⅢⅢ ⅢⅢ ⅢⅢ ⅢⅢ	

Make a pictogram of this information. Use each figure (♀) to show 10 pupils.

Exercise 13

Here is a tally chart of pupils attending a Youth Club.

day	tally	total
Monday	ⅢⅢ ⅢⅢ ⅢⅢ ⅢⅢ	
Tuesday	ⅢⅢ ⅢⅢ ⅢⅢ ⅢⅢ ⅢⅢ ⅢⅢ	
Wednesday	ⅢⅢ ⅢⅢ ⅢⅢ ⅢⅢ ⅢⅢ ⅢⅢ	
Thursday	ⅢⅢ ⅢⅢ ⅢⅢ ⅢⅢ ⅢⅢ	
Friday	ⅢⅢ ⅢⅢ ⅢⅢ	

Make a pictogram of this information.
Use each figure (♀) to show 10 pupils.
Use half the figure (♀) to show 5 pupils.

ar charts

Because pictograms often take a long time to draw, a bar chart can be used.
Here is the same information displayed on a pictogram and a bar chart.

Pupils attending activities at the Youth Club.

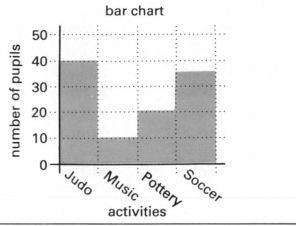

Exercise 14

Re-draw these pictographs as bar charts.

1.

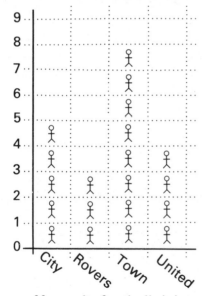

Survey of favourite football clubs

♀ = 1 pupil

2.

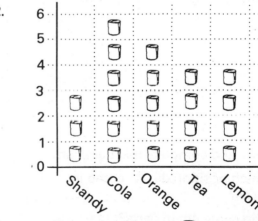

Survey of favourite drinks. = 1 drink

3.

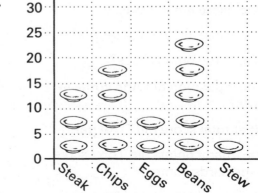

Survey of favourite foods. = 5 portions

Statistics at work

Exercise 15 Six children take part in a competition.
You can see their positions for each event in the four red boxes.

Points	Darts	Cards	Chess	Table Tennis
1st = 11 points	1st Mary	1st Errol	1st Tony	1st Ann
2nd = 9 points	2nd Tony	2nd Mary	2nd Ali	2nd Tony
3rd = 7 points	3rd Errol	3rd Ann	3rd Errol	3rd Mary
4th = 5 points	4th Ann	4th Ali	4th Mary	4th Errol
5th = 3 points	5th Ali	5th Fred	5th Ann	5th Ali
6th = 1 point	6th Fred	6th Tony	6th Fred	6th Fred

1. Copy and complete this table.

Name	Points	Position
Ali	3 + 5 + 9 + 3 = 20	
Ann		
Errol		
Fred		
Mary		
Tony		

2. Copy and complete this bar chart.

3. Which position did Ali come in chess?

4. Who came 4th in darts?

5. Who scored a total of 9 points?

6. Who came 6th in cards?

7. Who came 1st in table tennis?

8. Who came 5th in chess?

9. Which position did Ann come in darts?

10. What total score did Errol get?

11. How many points did Tony score in the cards contest?

12. Who scored 5 points in the darts contest?

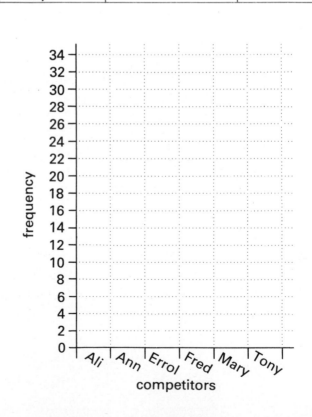

Patterns and tessellations

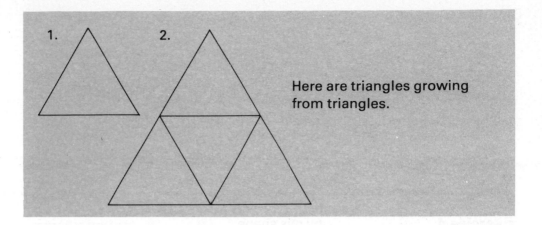

Here are triangles growing from triangles.

Exercise 1

1. Copy the triangle patterns into your book.
2. If the triangle pattern continued to grow, what would the next triangle pattern look like?
   Draw the next two triangle patterns in your book.
3. How many triangles make up   **a.** the first triangular pattern?
   b. the second triangular pattern?
   c. the third triangular pattern?
   d. the fourth triangular pattern?

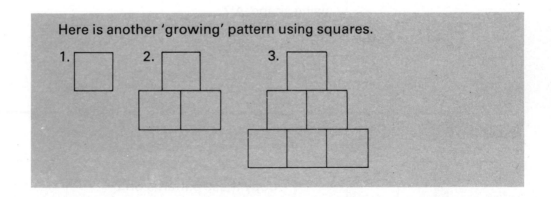

Here is another 'growing' pattern using squares.

Exercise 2

1. Copy the square patterns into your book.
2. Draw the next two square patterns.
3. How many squares make up   **a.** the first square pattern?
   b. the second square pattern?
   c. the third square pattern?
   d. the fourth square pattern?
   e. the fifth square pattern?

Exercise 3

Copy this triangle.

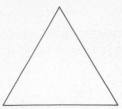

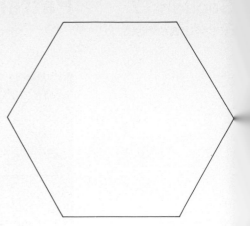

1. Can you make this shape using the triangle?
2. How many triangles are needed?

Exercise 4

Copy this shape.

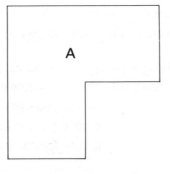

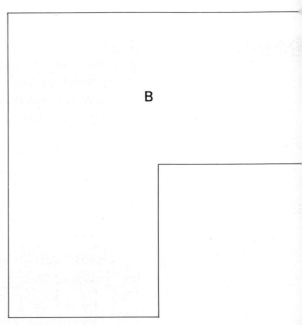

1. Can you make shape 'B' using shape 'A'?
2. How many shapes like 'A' are needed to make shape 'B'?

Exercise 5

1.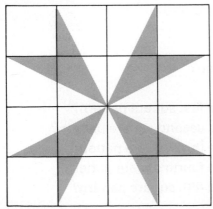

Copy these patterns onto squared paper.

Continue the pattern.

You will make other shapes as the pattern grows.

What other shapes are made?

2.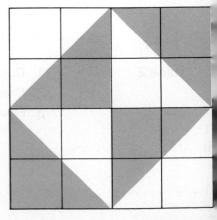

Tessellations

A tessellation is a pattern which is made by repeating a shape over and over again. The pattern must have no gaps or other shapes in it.

This is a tessellation of squares.

A shape can be moved round in the pattern.

Exercise 6

Which patterns below are tessellations?

1.

2.

3.

4.

5.

6.

7.

8.

Exercise 7

This is a tessellation of rectangles.

There are a lot of tessellation patterns that can be made using a rectangle.

Trace or copy this red rectangle and see how many different tessellation patterns you can make.

Exercise 8   Redraw or trace these shapes and see if they will tessellate.

Colour the finished patterns to make a wall display.

Exercise 9   Use one of the shapes below to make a tessellation.

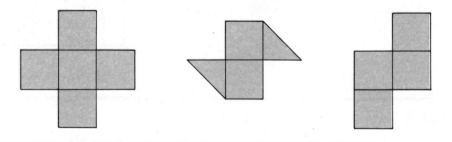

> Bricks form tessellations.
> Bathroom and kitchen tiles form a tessellation.
> Think of anything else which will tessellate.

Exercise 10   1. Draw tessellations that you find around you. Which shapes do the tessellations use?
2. Design a set of bathroom tiles.

Review 4

. Number work

Write out the answers.

a. $4 + 5 + 3 = *$ **b.** $5 + 6 + 2 = *$ **c.** $8 + 6 + 7 = *$

d. $9 + 6 + 5 = *$ **e.** $12 + 3 + 4 = *$ **f.** $15 + 8 + 5 = *$

g.	h.	i.	j.
12	20	23	31
14	6	17	9
+3	+13	+8	+22
___	___	___	___

k.	l.	m.	n.
36	42	51	60
−14	−32	−12	−33
___	___	___	___

o.	p.	q.	r.
40	46	51	68
−19	−37	−23	−18
___	___	___	___

s. $4 \times 5 = *$ **t.** $5 \times 6 = *$ **u.** $3 \times 6 = *$ **v.** $7 \times 3 = *$

w. $3\overline{)9}$ **x.** $2\overline{)12}$ **y.** $2\overline{)24}$ **z.** $5\overline{)25}$

. Patterns

See if you can draw this pattern in your book.
Use colour or shading to make it look interesting.

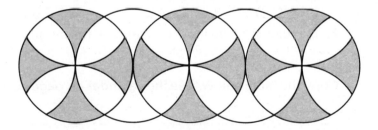

C. Angles

1. Copy and complete these sentences.
   a. A right angle is_____°.
   b. Acute angles are more than _____° and less than_____°.
   c. Obtuse angles are more than_____° and less than_____°.
   d. Reflex angles are more than_____° and less than _____°.
   e. A straight line is an angle of_____°.

2. Copy these drawings in your book. Make the lines longer.
   Measure each angle. Is it acute, obtuse, reflex or a right angle?

   a.   b.   c.   d.   e.

D. Decimals

1. Which of the drawings below show 0·1 coloured?

a.

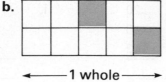

←————————1 whole————————→

b.

←———1 whole———→

c.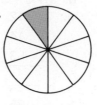

1 whole

d.

1 whole

2. Which of these numbers is the same as two units and seven tenths?
   a. 3·6   **b.** 1·9   **c.** 2·7   **d.** 7·2

3. Which of these numbers is bigger than 6·4?
   a. 5·9   **b.** 6·8   **c.** 0.9   **d.** 6.3

4. Which of these numbers is smaller than 0·6?
   a. 0·3   **b.** 0·7   **c.** 1·0   **d.** 8·7

5. Which of these numbers shows 5 tenths?
   a. 1·4   **b.** 2·7   **c.** 5·4   **d.** 0·5

6. Copy these expressions. Use the > or the < sign to make the expressions true.
   a. 6·1___6·0   **b.** 0·9___1·0   **c.** 0·5___0·1   **d.** 7·0___0·7
   e. 3·3___7·2   **f.** 4·4___0·4   **g.** 4·0___0·4   **h.** 0·7___7·0

7. If a ruler is 10 cm long, one tenth of the length of the ruler will be ___cm.

E. Time

Copy the table below. Use the calendar on page 66 to help to fill in the table.

Month with 28 days	
Months with 30 days	
Months with 31 days	

Statistics

1. Here is a drawing of Robert Wadlow, the world's tallest man.
The bar chart tells you how tall he was during his younger years.

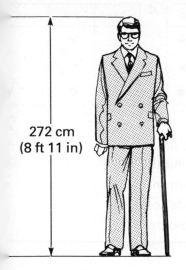

272 cm
(8 ft 11 in)

Robert
Wadlow

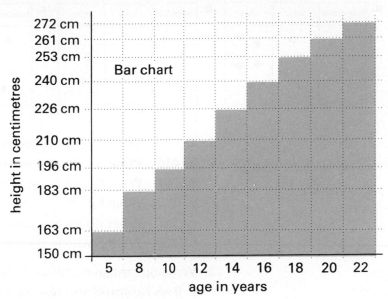

Bar chart

height in centimetres

272 cm
261 cm
253 cm
240 cm
226 cm
210 cm
196 cm
183 cm
163 cm
150 cm

5 8 10 12 14 16 18 20 22
age in years

a. How tall was Wadlow at 8 years of age?

b. How tall was Wadlow at 14 years of age?

c. How tall was Wadlow at 10 years of age?

d. How tall was Wadlow at 18 years of age?

e. How tall was Wadlow at 20 years of age?.

f. How old was he when his height was 210 cm?

g. How old was he when his height was 240 cm?

h. How old was he when his height was 163 cm?

2. Record these dice scores on a tally chart and on a bar chart.

| 4 – 5 – 1 – 3 – 1 – 6 – 4 – 3 – 1 – 5 – 6 |
| 4 – 1 – 2 – 5 – 3 – 6 – 4 – 4 – 4 – 2 – 5 |
| 1 – 4 – 5 – 2 – 1 – 3 – 6 – 1 – 5 – 4 – 4 |

number	tally	total
1		
2		
3		
4		
5		
6		

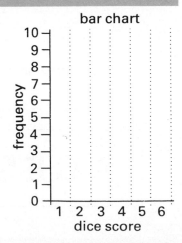

bar chart

frequency

10
9
8
7
6
5
4
3
2
1
0

1 2 3 4 5 6
dice score

G. Base ten

1. What numbers are shown here?

a. hundreds	tens	units
▲	o o o	● ● ● ● ●

b. hundreds	tens	units
◢ ◢ ◢ ◣	o	● ● ●

c. hundreds	tens	units
▲ ◢ ◣		● ● ● ● ● ●

d. hundreds	tens	units
◢	o o o o	

2. a. 125 What figure is shown in the tens column?

    b. 416 What figure is shown in the units column?

    c. 264 What figure is shown in the tens column?

    d. 359 What figure is shown in the hundreds column?

    e. 247 What figure is shown in the units column?

    f. 420 What figure is shown in the hundreds column?

3. Write out these numbers in figures.

    a. Two hundred and twenty-six.

    b. Six hundred and sixty-five.

    c. One hundred and fifty-one.

    d. Four hundred and seventy.

    e. Eight hundred and five.

H. Sorting

1. Sort these shapes into three groups.

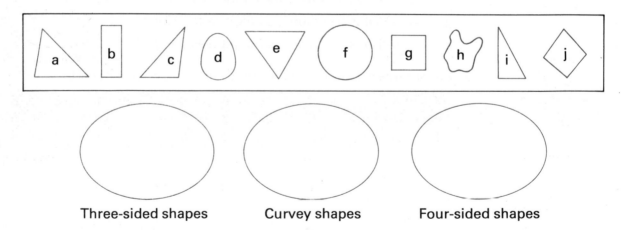

Three-sided shapes   Curvey shapes   Four-sided shapes

2. Sort these letters into three groups.

A C E S D P L O Q M R Y Z K U W

    a. Straight letters   **b.** Curved letters   **c.** Curved and straight letters

Street maths

Targets

Exercise 1

Here are the targets from a target shooting competition.

Each person had 5 shots.

1.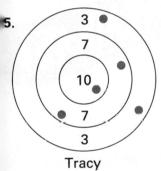
Jane

2.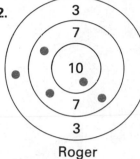
Roger

3.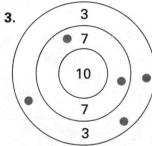
Peter

4.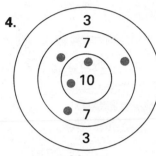
Mary

5.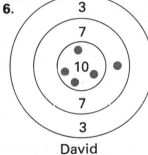
Tracy

6.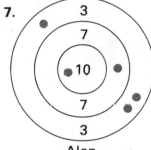
David

7.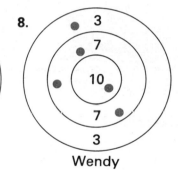
Alan

8. Wendy

1. Copy and complete the table.

2. Who scored 23 points?

3. Who came second?

4. How many points did Wendy score?

5. Who came sixth?

6. Who scored 36 points.?

7. How many points did Alan score?

8. Who came first?

9. Who hit most tens on the target?

10. Who scored no tens on the target?

11. Who scored 34 points?

Name	Points	Position
Jane	3 + 3 + 10 + 10 + 10 = 36	

Fun Fair

Exercise 2

1. How much would a go on the Hoop-La and the Bumper Cars cost together?
2. How much would it cost to visit the Fortune Teller and the Horror House?
3. How much would it cost to go on the Big Dipper and the Hoop-La?
4. How much would two goes on the Bumper Cars cost?
5. How much would two visits to the Hall of Mirrors cost?
6. How many candy flosses would you get for 50p?
7. How much would two goes on the Ghost Train cost?
8. How much would two goes on the Boating Lake cost?
9. How much would three goes on the Bumper Cars cost?
10. How much would it cost to go on the Big Dipper and the Ghost Train?
11. How much would three visits to the Hall of Mirrors cost?
12. How much change from £1 would you get if you had a go on the Big Dipper?
13. How much change from £1 would you get if you bought a candy floss?
14. How much change from £1 would you get if you went on the Ghost Train?
15. If you had £1.20, how would you spend it at this fair?

Pinball Wizard

Exercise 3

In each pinball game, ★ means a score.

1. Work out the score for each go.
2. What is the total score after four goes?

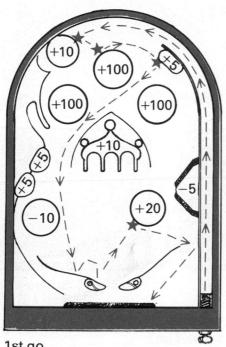

1st go

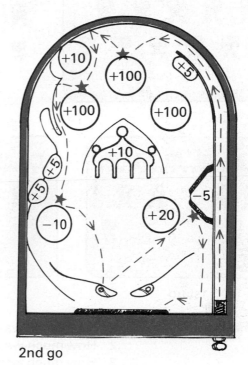

2nd go

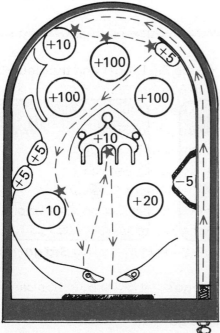

3rd go

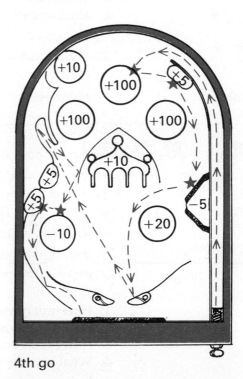

4th go

Space Invader

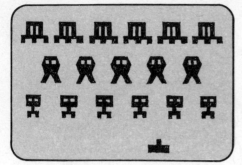

Here is a space invader screen before the game starts.

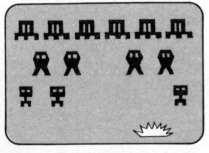 = 120 points

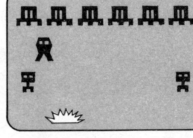 = 50 points

 = 25 points

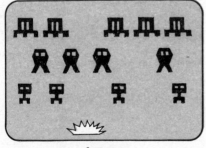

1st go

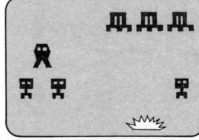

2nd go

3rd go

Exercise 4

1. How much was scored on the first go?
2. How much was scored on the second go?
3. How much was scored on the third go?
4. What was the total score?

Exercise 5   A new sheet

1st go

2nd go

3rd go

1. How much was scored on the first go?
2. How much was scored on the second go?
3. How much was scored on the third go?
4. What was the total score?